SURVIVAL FOR AIRCREW

To Trevor

Survival for Aircrew

SARAH-JANE PREW

Illustrations by Simon Austin

Ashgate

Aldershot • Brookfield USA • Singapore • Sydney

Published by
Ashgate Publishing Ltd
Gower House
Croft Road
Aldershot
Hants GU11 3HR
England

Ashgate Publishing Company
Old Post Road
Brookfield
Vermont 05036
USA

British Library Cataloguing in Publication Data
Prew, Sarah-Jane
    Survival for aircrew
    1. Survival after airplane accidents
    2. Aircraft accidents - Human factors 3. Flight crews -
    Training of
    I. Title
    613.6'9'024'6291

Library of Congress Catalog Card Number: 99-72608

ISBN 1 84014 521 8

Printed and bound by MPG Books Ltd, Bodmin, Cornwall

Contents

List of Illustrations		*vii*
Preface		*ix*
Acknowledgements		*x*
1	Survival - An Introduction	1
2	The Crew Role - Leadership in Survival	5
	Sea Survival	
3	Water Survival - It Can Happen to You!	8
4	Ditching - How to Prepare	14
5	Underwater Escape	17
6	In the Water	21
7	Liferaft Management	31
8	Liferaft Design and Equipment	38
9	First Aid at Sea	49
10	Signalling and Rescue at Sea	56
	Land Survival	
11	Introduction - Land Survival	70
12	The Desert	73
13	The Jungle	89
14	The Arctic	97
15	Lighting Fires	107

16   First Aid on Land   110

17   Signalling and Rescue on Land   119

18   Conclusion   124

Sources for Equipment and Training   *125*
Bibliography   *128*
Index   *129*

List of Illustrations

The ditched Royal Air Force Nimrod 10
An Air China Boeing 747 in the harbour at Hong Kong Airport 11
Piel Emeraude, ditched in the Irish Sea 13
Training for underwater escape 18
Infant life preserver 22
Conserving body heat in the water 25
A group huddling together 26
Flotation with a seat cushion 27
Boarding a raft 32
Rafts tied in a line 34
Different shape rafts in the wind 39
Cabin crew fasten the self-erecting canopy 40
Raft with self-erecting inflatable canopy 41
Raft with one tube 42
Raft with two tubes 42
Capsized raft with inflatable canopy 42
The drogue on a liferaft 43
A raft with all the ideal features 44
Administering Cardio Pulmonary Resuscitation 50
The recovery position 51
The five sections of the COSPAS-SARSAT system 60
Using a signalling mirror 62
SEE/RESCUE streamer in the water 65
SAR crews rescuing an injured casualty 68
Use the outside of the aircraft as a source of shade 75
A simple shelter 76
A liferaft leaning against a tree can provide shade 77
Making an Arab head dress is simple 78
A distillery 80
A transpiration bag 81
A solar still 82
A simply constructed shelter 85
An example of lashing 85
A cave can make an excellent shelter in the desert 86

Bent down branches can provide shelter	95
Snow shoes	99
A snow hole	101
A snow trench	102
A shelter using branches and leaves	104
Lighting the fire on a base of logs	108
A candle can be an invaluable tool for fire lighting	109
Apply pressure and elevate	112
A SAM splint	113
If you suspect a chest injury	114
Use whatever is around you to attract attention	122
A SEE/RESCUE streamer	123

Preface

Survival for Aircrew has been written as an instructional tool for all those involved in flying over wilderness areas or water. It places special emphasis on the crew role in survival situations, whether the reader is the pilot of a Cessna with three passengers on-board, or the Purser of an airline flight.

The book is aimed to teach survival techniques for crew and passengers who may find themselves in a survival situation without warning. They are likely to be unprepared. It is not supposed to be a comprehensive survival guide such that those venturing into wilderness regions intentionally may wish to purchase.

Rather, it leads aircrew through survival techniques that are within their grasp. It does not require specialist knowledge or fancy equipment, but advises on how each member of the team can improvise with what they have, or with what nature can provide, to pull through and survive.

There is no chapter on snaring animals, for example, for two reasons. The first is that, except in the very cold regions, food is not considered a priority to survival. Humans can live without it long enough to be rescued from an aircraft crash and in any case, food requires more water to digest than is likely to be available in a survival situation.

The second reason is that it is considered beyond the scope of those in an air crash and survival situation to be able to catch animals safely. There will be other priorities that will absorb all time and energy. The dangers involved with attacking wild animals is one extra threat that survivors can do without.

Similarly, it is considered beyond the scope of pilots and cabin crew undertaking normal operations to learn and recognise edible plants, merely from reading this book.

Survival for Aircrew covers what crew members need to know. It does not waste the readers' time and money on information that is not relevant to their situation.

It must be made clear, however, that this book does not eliminate the need for thorough practical training. Survival for aircrew simply provides a comprehensive instructional tool on which to base other such experiences.

Acknowledgements

As ever with a book such as this, there are many people who contribute towards its existence and I certainly have my fair share of acknowledgements to make. The following have helped in numerous ways, from providing information, photographs and advice, to allowing me to attend survival training courses and operational Search and Rescue missions.

Whatever the nature of the help received, I can only thank all those who have aided me in the writing of Survival for Aircrew. If anyone has been left out of the list, my sincere apologies - it was certainly not intentional.

There are a number of organisations throughout the world who provide survival training of various kinds. Many of these institutions have been kind enough to let me attend courses, many in the wilderness.

First and foremost I must thank the Warsash Maritime Centre, part of the Southampton Institute, whose professional survival instructors have welcomed me on a myriad of courses including sea survival, underwater 'dunker' training and helicopter winching courses. Thanks for these opportunities must also extend to Andark Diving and to the UK Coastguard, and in particular to the Chief Pilot and crew of HM Coastguard SAR helicopter BDIJ, based at Lee-on-Solent.

Other organisers of survival courses in which I participated include the UK Department of Transport, Brian Horner at LTR Training Systems Inc and Ken Burton at S.T.A.R.K. Survival.

My thanks must also extend to the US Coast Guard, and in particular to Lt Cdr Paul Steward in Washington and to Lt Cdr Jennifer Lay, Senior Rescue Controller at Miami, who allowed me to visit the control centre and observe the team at work. Thanks also to the team at Opa Locka Coast Guard Station who made me so welcome during my visit.

Special thanks go to the RAF School of Combat Survival at RAF St Mawgan in Cornwall and also to Steven Callahan, the yachtsman who spent 76 days alone in a liferaft after his yacht capsized. His stories of survival should be an inspiration to us all.

There are other institutions to mention that have taken the time to help and advise me, or again, have supplied photographs. These include the Royal National Lifeboat Institution (RNLI), the Civil Aeromedical Institute (CAMI), the UK Air Accident Investigation Branch (AAIB), the UK Civil Aviation Authority (CAA), the National Transportation Safety Board (NTSB) and the

Federal Aviation Administration (FAA). Thanks must also go to Boeing and Airbus.

Many airlines have allowed me to attend their cabin crew training courses and many have supplied photographs. Among these I must mention, with particular thanks, Terry King at British Airways and Simon Gray at British Mediterranean Airways. In addition, my thanks extend to the following airlines: SAS, Malaysia Airlines, Delta Air Lines, Emirates Airline, Thai Airways International, Lufthansa, Virgin Atlantic Airways and EVA Air.

The other group of companies that must be mentioned are the equipment manufacturers. Two in particular must receive special thanks because they have been extremely patient in answering my repeated questions on a whole host of relevant topics. These are Winslow Life Raft Company from Florida and HR Smith Group of Companies, based in Leominster, UK.

Others also deserve a mention, however, and I thank them for their time. These include BF Goodrich, Beaufort, RFD, Eastern Aero Marine, Icarus, SEE/RESCUE, The Seaberg Company and Pains Wessex.

I must also mention my parents. My father contributed to the artwork while my mother offered her proofing and administrative services.

There are two other people without whom the book would not be what it is. The first is Simon Austin, who provided me with all the excellent black and white sketch drawings throughout. The second is Trevor Nash, whose diligence in reading through the 'finished' manuscript brought to light all those editorial howlers that would otherwise now be adorning the page. His support throughout the project made the whole thing possible.

1 Survival - An Introduction

Survival is the art of staying alive and for the purposes of this book it is about staying alive in extreme climatic conditions. It is the ability to combine our technological knowledge and experience with the old natural skills that are so quickly being eroded away by our modern lifestyles. The right psychological attitude is essential and is man's biggest survival asset; knowledge and prior training is the next most valuable tool.

Survival situations occurring from aircraft accidents happen suddenly and with relatively little warning. It is unlikely that anyone boarding that aircraft at the beginning of the journey would have given any thought to the possibility that they could end up in a life-threatening survival situation.

These events happen when people are least expecting them. They are on the aircraft for a reason; they have their trip planned out and their mind is on many things. When something then goes wrong and people are left struggling for their lives in the desert or in the oceans, it is such a complete shock that they do not know how to begin to cope.

Anyone who succeeds in this task will do so by making themselves a survivor. They will stay calm and think clearly. They will be positive and will see opportunity in everything. They will make the most of the equipment that they have and each item will suddenly assume a dozen uses. If they do not have equipment, they will not see themselves as unequipped, merely that they must turn to the natural resources around them, of which there are many. They will learn, very quickly, to improvise and will become instantly resourceful. They will appreciate that although their surroundings are trying to kill them, their surroundings are all they have that will keep them alive.

There will be many dangers threatening life in a survival situation. These will include the freezing cold, the searing heat, the relentless sun, the sea, wind, lack of water, animals and insects, hunger and injury. These are physical dangers whose presence the survivor will have no control over.

Aside from these, there will be many more dangers, most of which will be conjured up by those in the survival situation. These are the psychological dangers and can include panic, fear, fatigue, boredom, a feeling of isolation and a lack of will to live.

Psychological factors can make death a certainty. It is very important that these barriers are addressed and overcome, or not allowed to build up in the first place. This applies first and foremost to the crew but it also applies to the passengers. It is the crew's job to do all they can to prevent panic from spreading and negative feeling from destroying the survival effort.

The recognition phase is the point at which the reality of the situation sets in. This is when the brain realises that something is wrong and, worse than this, sees all the implications of that situation. The greater the threat that the brain perceives, such as a life threatening situation as opposed to a non-life threatening situation, the greater the stress felt by the body. This stress can produce negative effects on our thinking patterns and actions. Certain people, especially those who do not have very good coping mechanisms, suffer particularly adverse reactions to stress.

The recognition phase is most likely to set in once the panic of the crash landing and evacuation is over. The relief that you have survived an air crash will quickly be overtaken by the more deep-rooted panic that you are in a hostile environment and not likely to be rescued in a hurry.

There are three recognised categories of survivors; the leaders, the followers and those who are negative about the whole event. The leaders form about 25% of the group. These are the natural leaders and the most unexpected people can come to the fore in such situations. While it is hoped that the crew of an aircraft will come into this category having, hopefully, been chosen for their role as leaders, this may not always be the case. It is important that crew do assert themselves as leaders, however, because they may be the only ones with the relevant knowledge to keep everyone alive.

The next category, the followers, represent about 50% of the population. These are the people who will obey and do everything that is required of them but they will wait until told to do so. They are not obstructive in any way and want to help but will not jump up and lead.

The third category, the final 25%, comprises those who are totally obstructive and who will damage the survival process if they can. They remain negative and refuse to believe that anything anyone tries to do will do them any good whatsoever.

The basic rule in survival is 'help yourself first'. Do not sit back and rely on someone else - that person may never materialise. As crew, if you do not help yourself, no-one else will and your passengers will be relying on you to help them.

Your survival will depend upon a will to live. A real will to live will be stronger than any dangers trying to kill you. Fear, for example, can act in two ways. It can cause you to panic, making you an indecisive, gibbering wreck or it can sharpen your instincts and make you think all the clearer, forcing you towards helpful action.

As crewmembers, it is up to you to get yourselves and passengers through this. They are reliant on you for knowledge, guidance and support. Except in occasional circumstances, your passengers will know nothing other than what you tell them. It is up to you to help them save their own lives.

Do not hold post-mortems once the crash has occurred; these will do no good whatsoever. Instead, keep together, assess the situation and concentrate on surviving. You must be decisive, able to improvise and able to adapt to any situation. You must maintain hope, stay calm and you must have patience. You must be able to take hardship to its limits. Knowledge in the subject and awareness of the very worst that can happen to you helps you accept this hardship.

One of the best ways of looking at survival is to take the advice of those that have been through such situations. Steven Callahan, who spent 76 days alone in a life raft after his yacht sank, maintains that you must 'see your predicament as nothing more than a continuation of your journey'.

This is important. If you feel hard done by, you will be less able to get down to work and do what must be done to keep people alive. Accept the situation and live with it. Use whatever you have and be thankful for it. Think for yourself and make the most of each tiny thing. It is important to realise the richness of your surroundings; even though hostile, there are so many things in the wilderness that you can use. The survival skill is to recognise this richness for what it is, not wish you had your latest technological gadget with you.

Steven Callahan backs this up. 'Although the sea was my greatest enemy, it was also my greatest ally....the sea is indifferent but her richness allowed me to survive.'

Everything in survival is paradoxical. Every positive decision will have a negative side to it. It is just down to the individual to get on with it and make the most of everything. Ultimately, you have a choice - you either survive or you curl up and die.

Managing your own personality is the most crucial aspect. Know how you manage under stress. Know how much you can take and what effect it has on you. The rational self must take command over the pain, the fear, the thirst, the hunger and the will to give it all up. Through all these extremely strong negative feelings must come just one thing - the will to live and through this the energy to get up and carry on.

Do not underestimate the effects of the mind in a situation such as this. To use Steven Callahan as an example again, he found it increasingly difficult to kill the fish around him, his only source of food, because they were his only companions and had become his friends.

To use another example, one of five yachtsmen whose boat sank, was convinced after a couple of days that he could see land close to the raft. He was so sure about this that he dived into the shark infested sea and swam away from the raft. He was never seen again. His brain was playing particularly nasty tricks with him.

Yet another example can be seen by looking at a survivor of a Cessna that crashed in the Sierra Nevada mountains in 1976. A woman, trying to reach help by walking through the snow and ice, saw a row of houses. She saw people and even heard children laughing and shouting. There was nothing there, in reality, but snow capped mountains and rocks. When she realised that the view was nothing but a figment of her imagination, her morale was severely affected. She had also wasted a lot of energy in trying to reach what was never there.

Because the psychological dangers are so powerful and threatening, it is important first to recognise them and secondly to control them, both in yourself and then in others. Keeping everyone busy is the key to this. If people are doing something, it keeps them active and keeps their mind from the worst. They may also feel that they are helping themselves and others to survive. A feeling of responsibility for the success of their plight can go a long way to increasing the will to live.

Have a roster and make sure people stick to the tasks on it. Do not have people working all the time; they need to be able to rest, especially as they will be tired, possibly injured, thirsty and hungry. If they have periods of activity and responsibility between rest, however, it will do them a great deal of good. It will also get the essential work done.

There are certain times when it is impractical to work; for example during the day in the desert. At these times, keep people's minds occupied when they are not asleep. You cannot afford for them to begin thinking too much about their plight. Positive thought may quickly turn to negative and this can be very damaging to all involved. Panic will spread like wild fire so do not let it build up in the first place.

There is one basic survival rule to follow: the priorities of survival, PROTECTION, LOCATION, WATER and FOOD. These terms are self explanatory, save for location. In survival terms, location means attracting attention and being rescued. Food can be disregarded in all but extreme cases. People can survive a long time without food and it is a commodity that can take a lot of energy to acquire, and a lot of water to digest. Unless you have enough water to give everyone about five litres a day, forget food.

The priorities will depend on the individual circumstances. The basic aim is to get rescued and remain alive to be rescued. This means that there are two priorities that should be working in conjunction: protection will ensure that everyone is protected from the immediate dangers, for example, the sea, wind, freezing cold or sun, allowing them to survive, while location will ensure a constant look-out for rescue.

Injuries may be a significant factor but it is no good worrying about these, however life-threatening, if exposure or drowning kill you first. Water is also a priority, but will not kill as quickly as climatic dangers so prioritise according to the threat to life.

There are a few other basic survival rules to follow. At sea, never allow people to drink sea water; it will kill quicker than thirst. NEVER give anyone alcohol, especially in cold conditions. Although it may appear to warm the body, all it is actually doing is dilating the blood vessels which causes the blood to flow around the body faster. You feel warm because the warm blood that was protecting the vital organs in the very cold body core, is now at the surface, where you can feel it in your extremities. The cold blood it has replaced is now close to the vital organs and the shock of this will kill. Alcohol is also dehydrating.

In summary, to survive, keep positive and never give up.

2   The Crew Role - Leadership in Survival

With the exception of the airline Captain and possibly the senior cabin crewmember, the crew of an aircraft will rarely have had any training in leadership. In a survival situation, however, the need for leadership and management skills will come to the fore, especially if the incident is prolonged.

So what is leadership and management? According to one famous military commander, Field Marshall Montgomery, leadership is, 'the will to dominate, together with the character which inspires confidence'. Management on the other hand, at least according to M. Gilbert Frost, 'is the art of directing human activities'.

Consider the leadership and management roles of the crew in the environment posed following an aircraft accident when help is not immediately at hand. Remember the old adage that, To Fail to Prepare, is to Prepare to Fail.

Imagine the crew's role in a survival situation. After the difficulties and stresses of an emergency landing and passenger evacuation, at the point where there would normally be someone else, such as the police or airport staff, to take over, your role in the survival situation is just beginning. Passengers, and indeed junior crew members, will be looking to the more senior crew members for help, support and leadership. This, despite the crews' own problems, possible injuries and emotional turmoil.

Success in survival depends on, among other things, knowledge and ability, the psychological condition of the individuals and on the management of the survival effort. All these factors are dependant on the crew and on their ability to manage and lead.

Consider the survival situation. Chaos will reign. There will be injuries to deal with, shelter to find, water to acquire, fires to build and try to light, rescue attempts to look out for and attract. People will be in pain, in shock, panicking and confused. As well as managing this chaotic situation and setting priorities, the crew member must utilise resources to prevent further injury and turn chaos into order.

The role of the leader in these situations is multi-faceted. The leader will make the decisions about what has to be done and will prioritise tasks. They will control the survival effort. They will be required to maximise the resources available, to determine improvisations and to maintain the safety and efficiency of the group. Perhaps most importantly, they will be required to build and maintain morale.

Morale is certainly the key to survival. It is a state of mind. It is an intangible force that will move a whole group of people to give their last grain of strength to achieve something without counting the cost to themselves. The leader's role is to build and sustain morale and then direct it to the common goal of survival and well being for the group.

The maintenance of good morale in the survival group demands that the following are achieved:

* Leadership.
* Discipline.
* Comradeship.
* Self-Respect.
* Devotion to a Cause (ie Surviving).

It is the crew's role to take care of the passengers. They will be totally reliant on you for guidance. Except in very rare circumstances, you will be the only people in the party who have enough knowledge to be able to sustain the survival effort successfully. The onus is, fairly and squarely, on your shoulders. This responsibility is a moral one which, unfortunately, airlines do not formally address. The number of airlines that undertake formal leadership training is minimal.

In a survival situation, it is imperative that your leadership skills, and your knowledge of survival skills, show through. Crew must be able to control a very difficult situation and the people involved within that situation. They must be able to think clearly and organise all the tasks to be done, in order of priority. They must be able to delegate and impart requirements clearly. Crew must be able to instil confidence, respect and authority in the whole group.

Even more important than the relationship between crewmember and passenger in these situations is the ability for the crew to work together. The most senior crewmember should take ultimate command and should delegate down among the crew, taking into account individual strengths and weaknesses. The crew must be able to bond closely together and work as a team. Each crewmember, however, has a responsibility to the passengers.

In considering the use of all available resources, the crew must not forget the passengers. The chances are that on a long-haul flight for example, some of the passengers might be military or ex-military personnel that might have a particular experience in survival. Others may be medically trained. These personnel must be utilised without handing over authority. The crew has both moral and legal responsibility to their passengers. If the latter is in doubt, consider some of the litigation that could follow an accident in which it was proved aircrew caused death and injury to passengers through a failure in both knowledge and leadership skills.

The tasks in a survival situation are varied and enormous. The most important asset in any leader is to be able to oversee the management of all the necessary tasks and avoid wasting time on those that are not essential. The leader must listen to advice, assess the situation, plan resource utilisation and then work out a strategy before supervising its implementation.

Having examined some of the tasks demanded in a survival situation, questions emerge such as, what is a leader, can leadership be taught and will the crew have such a person in the team to lead them through the immense adversities posed by the survival situation?

In 1934, the Lord Bishop of Durham gave the Walker Trust Lecture on Leadership at the University of St Andrews, in Scotland. In his lecture, the Bishop high-lighted the whole question of whether leaders are born or develop. Although given over 60 years ago, the point made is as valid today as it was when it was written!

'It is the fact that some men possess an inbred superiority which gives them a dominating influence over their contemporaries, and marks them out unmistakably for leadership. This phenomenon is as certain as it is mysterious. It is apparent in every association of human beings in every variety of circumstances and on every plane of culture. In a school among boys, in a college among the students, in a factory, shipyard, or mine among workmen, as certainly as in the Church and in the Nation, there are those who, with an assured and unquestioned title, take the leading place, and shape the general conduct.'

Leadership skills, which are inherent in many, can be brought out and developed through training and it is undoubtedly this area of airline operations that is not taught sufficiently to both flight deck and cabin crew members. Taught or not, aircrew will be expected to both be survival experts and leaders should the survival situation arise.

Although leadership is a complex topic to many, the leader relies on certain skills to command and direct a situation. In no particular order, the leader should possess the following skills:

* Knowledge.
* Be a Good Manager.
* Be a Good Communicator.
* Be Approachable.
* Be Fair.
* Maintain a Sense of Humour.
* Show Self-Discipline.
* Set an Example to Others.
* Be Honest.

Over the following chapters, a number of different survival environments will be discussed along with the skills involved to allow the reader to prevail in them. This chapter will hopefully provide a backdrop in front of which the other chapters will be read. To reiterate, as well as personal survival skills, passengers will be looking to you, as a crew member, to assist them in completing the rest of their journey to safety.

Sea Survival

3   Water Survival - It Can Happen to You!

On May 2nd 1970, an Overseas National Airways McDonnell Douglas DC-9-33F, operating on behalf of ALM Dutch Antillean Airlines, ditched in the sea in the West Indies.

Having taken off from New York, enroute to St Maarten, West Indies, the flight was diverted twice due to bad weather. The Flight Crew began to get erratic readings from their fuel gauges, which showed next to no fuel and the decision was made to make preparatory plans in case a ditching became necessary. The Purser was called to the flight deck and was told to instruct the passengers to don their lifejackets and prepare the cabin. It was not made clear to the Purser that a ditching was imminent at this stage; merely that it was a possibility.

The Purser made the announcement to the passengers and the other two cabin crewmembers helped passengers with life jackets. Many had problems getting the lifejackets out of their pouches while others had problems putting them on. The flight engineer came back to help the cabin crew.

Once the fuel gauges showed that the engines were about to fail, the Captain levelled the aircraft off and positioned directly over the crest of a swell. He lowered 15 degrees of flap when about 20 feet above the water and allowed the airspeed to decrease. He lowered full flap as the low fuel pressure warning lights began to flicker. The engines flamed out.

The only warning the cabin got of the landing was the flashing Fasten Seat Belt and No Smoking signs. No previous agreement had been made to use this as a warning and so no-one was looking for it or understood its meaning. There was no communication from the flight deck after the Purser was given the first warning to prepare for ditching. The Captain flew the DC-9 onto the swell at about 90 knots.

In the Cabin

The Purser and flight engineer had removed one of the aircraft's five liferafts and put it by the galley door, ready to throw out of the aircraft after the impact. Another cabin crewmember was securing things in the cabin at the same time. All three heard the engines cut out and saw the warning cabin lights flash on and off. They realised then, as late as this, that the landing was imminent.

Several passengers were still standing in the cabin, struggling with their lifejackets at the time. One cabin crewmember was still standing trying to help them. Five other passengers were seated but with no seatbelts on.

The Purser and Flight Engineer shouted at everyone to sit down while they themselves sat on the aft facing jumpseat. Neither had time to fasten their seatbelts. One of the cabin crew sat on the liferaft by the door with his back against the bulkhead.

Some passengers sat, braced and generally did anything they could to protect themselves. Others, though, seemed not to realise what was happening and just looked out of the window. Some, including one cabin crewmember, did not even sit down.

The impact was severe and those who were standing or who were not strapped in were flung the length of the cabin and killed. Six more were injured or killed when their seatbelt fastenings failed.

The Escape

The aircraft remained intact and all the crew, except for the standing cabin crewmember, survived uninjured. The port-side forward main door could not be opened but a crewmember did manage to open the galley loading door on the starboard side. The crew tried to retrieve the raft from the galley, where it was buried in debris but it inflated in the cabin, trapping one crewmember temporarily under it. All the crew in the galley could not then return to the cabin and so jumped into the sea.

By this time the DC-9 was partly awash. One passenger opened an overwing exit and most of the uninjured passengers on-board escaped through this. The Captain escaped through his windshield and swam back to open the port side overwing exit. He helped two passengers out and could not see anyone else.

The aircraft then sunk from view WITHIN 10 MINUTES OF HITTING THE WATER! No rafts were taken off the aircraft; only one slideraft was available and this only because it had broken loose.

Rescue

A Pan Am flight was diverted to the area and was able to confirm the position of the ditched aircraft on radar. Two helicopters arrived on the scene one and a half hours later and began to winch the survivors to safety. Thirty-seven people were rescued at this point. The rest had to wait for another hour before being winched to safety.

Of the 63 passengers and crew, 22 - including two children and one cabin crewmember - were missing. They all either died in the aircraft on impact or drowned while awaiting rescue.

******

The Water Survival section of this book aims to cover any situation in which aircrew may find themselves in trouble over water. Over 70% of the earth's surface is covered in water so it is imminently possible that this situation may occur. The next seven chapters are relevant to any aviator who at any time flies over water.

Although the main emphasis of this section of the book is to prepare crew for a full blown ditching at sea, experience has shown that water survival situations can happen far closer to home.

There have been numerous General Aviation ditchings. Unfortunately they occur all too frequently and the most common cause is fuel exhaustion. Larger aircraft, and typically airliners, are not so prone to ditching. Many in the aviation field dismiss the idea of sea survival for airline aircrew altogether because they feel that even if it did happen, the chances of survival are nil.

Although extremely uncommon, there are examples that prove these cynics wrong. The Dutch Antillean DC-9 aircraft, described above, that ditched in the Caribbean is an example of a full blown ditching in which there were survivors. Another example includes the Ethiopian Airlines Boeing 767 which was forced to ditch off the coast of the Comoros Islands in 1996 due to fuel exhaustion after hijack demands sent the aircraft beyond its range. A few years before, a Royal Air Force Nimrod aircraft ditched in a Loch in Scotland. All on-board survived.

(With thanks to RAF Lossiemouth)

The ditched Royal Air Force Nimrod.

These examples show that it is possible to ditch a large passenger aircraft and survive. All these incidents happened in large expanses of water. This is water survival at its extreme but having some survival knowledge could save the lives of passengers and crew in aircraft that are not involved in anything like so dramatic an incident. Again there are examples.

The Air China Boeing 747 that skidded off the runway into the harbour at Hong Kong airport came to rest partially submerged. The aircraft's occupants had to evacuate into rafts and await rescue. Another even less obvious example involved the Air Florida Boeing 737 that failed to climb after take-off from Washington DC in the United States. It smashed into a bridge over the Potomac River, leaving five passengers and one crewmember clinging to wreckage in frozen water. Although the rescue services were on the scene instantly, they were unable to reach the survivors until a helicopter could be summoned; the ice prevented the launch of boats but was not strong enough to allow rescuers to walk over it.

The outlines of these incidents are intended to illustrate how easily a water survival situation can occur; to prove that people can survive such incidents and to warn that rescue is not always as quick and simple as many imagine.

An Air China Boeing 747 in the harbour at Hong Kong Airport.

A flight crew could be forced to ditch an aircraft for many reasons. One of the most common, as we have already seen, is fuel exhaustion. This is the cause of most General Aviation ditchings; it was the ultimate reason for the Ethiopian Airlines ditching. Although a cause, it is much easier to land an aircraft on water under power so it is uncommon for a landing to take place once power is lost. The imminent fuel exhaustion will force the decision to ditch but the flight crew will hope to get the aircraft down before this happens.

A pilot may be forced to ditch if the aircraft has suffered a severe mechanical problem that will not enable the aircraft to reach land. An in-flight fire is another very realistic reason. As with the Ethiopian Airlines incident, a hijack was the cause of the fuel exhaustion. A similar act of hostility towards an aircraft could, in extreme circumstances, force a ditching.

As in the case of the Air China and Air Florida incidents, these were no more than common take-off and landing errors; the most common phase of flight for any incident to occur. Should they happen near water, then the occupants of the aircraft could so easily find themselves having to survive in water.

As the above examples show, there is a fair chance of surviving these incidents. The chances of getting the aircraft down on the water successfully depend on many factors, including aircraft type, wind speed and most of all, surface conditions. The chances of the occupants evacuating the aircraft once it is on the water also depend on many factors. These include the nature of the impact and the injuries sustained, the consequent structural state of the aircraft, the outside conditions, the knowledge of the occupants or crew and their success in controlling passenger behaviour.

Once outside the aircraft, the occupants are in a survival situation and the chances of surviving this are down to two simple factors: the survival knowledge of the crew and the equipment available.

So there are three stages to a successful ditching. First of all it is essential to get the aircraft down as safely as possible. This is obviously the role of the flight crew. It is up to other occupants, however, to ensure that their chances of survival are as great as possible and this includes bracing appropriately, stowing all loose luggage, being prepared and having lifejackets on, but not inflated.

An aircraft will typically hit the water twice; the first time as the tail hits and the second when the nose hits. The second impact will be the greatest and can be surprisingly hard. A successful landing will leave at least some occupants alive and able to evacuate the aircraft.

Once the aircraft is on the water and has stopped moving, the evacuation takes place. This could take the form of an ordered evacuation into inflated rafts; it could mean everyone jumping into the water from every exit or could even mean people having to escape from an inverted, water filled aircraft. Time is of the essence, however people escape. A successful evacuation will see at least some occupants getting out of the ditched aircraft alive.

Once the evacuation has taken place, the survival then depends on everyone's ability to keep themselves alive at sea, whether in the water or, hopefully, in a raft.

Each of the three stages may take lives. It is generally understood, though, that over 80% of controlled ditchings are carried out with few, if any injuries. Half of these people, however, die before help arrives. This illustrates quite clearly that the human failure is in the sea survival part of the process and that lack of knowledge in this area can be fatal.

In a sea survival situation, to be a survivor you have to live until you are rescued. This could take any length of time from 20 minutes, if you are in a similar situation to the occupants of the Air China aircraft in Hong Kong harbour, to three days if the aircraft ditches, say, mid-Pacific. Never underestimate survival times. Assume you are going to have to wait a long time and you will prepare yourself for this; your chances will be greatly improved. Assume that you will be picked up very quickly and each hour that passes without rescue will reduce morale.

(Photo via John Thorpe)

Piel Emeraude, ditched in the Irish Sea.

The emphasis of the next seven chapters is to teach people to survive in water and to get themselves rescued. The chapters on water survival cover these dangers, preparation for ditching, protection against the dangers while in the water, the protection a raft can offer, raft management and one of the most important aspects of any survival situation, signalling and rescue.

4 Ditching - How to Prepare

Preparation is very important in a ditching. All aircraft occupants should be given a briefing before the flight. This should cover the safety features of the aircraft, including location and use of lifejackets if the flight is over water, together with where the exits are. If a ditching then becomes inevitable, this briefing needs to be reinforced so all on-board know what they are doing.

In General Aviation aircraft, the onus is on the Captain to give the briefing. In passenger aircraft, the cabin crew will fulfil this role. The onus is still on the flight crew, however, to warn the cabin crew and passengers that there is a problem in plenty of time. Very naturally, the flight deck crew are aware of the situation as soon as it arises. Everyone else is reliant on being told. Communication between the flight deck and cabin crew, therefore, is vital.

Referring back to the Dutch Antillean DC-9 aircraft. One of the worst mistakes here was the lack of information and warning from the flight deck about the imminent landing on water. Although current practice demands Crew Resource Management (CRM) training which reduces incidents such as this enormously, it must be remembered that having to ditch an aircraft is something most pilots would never have thought they would have to do and it is important that their attention is not focused so much on what they are doing in terms of flying the aircraft, that they forget to inform the cabin crew of their intentions.

In the majority of cases, there should be enough warning of a ditching for the occupants or cabin crew to prepare. This phase of activity should include preparation for the impact and also for the ensuing survival situation.

Passengers must be briefed, if possible individually. This is important as they can so easily jeopardise their own, and others', lives. As crew, you cannot rely on them remembering anything from the pre-flight briefing. The chances are, they were probably not listening anyway. You are now briefing them at the worst time possible. They will be frightened; maybe even panicking and you are expecting them to take in information. You must be patient and calming but very firm and you must explain everything very simply but clearly.

Many passengers may need help with their lifejackets. Those in light aircraft should be wearing them anyway when over water. Those in passenger aircraft will probably need to be shown where they are. It is important that lifejackets are fitted correctly and IT IS VITAL THAT THEY ARE NOT INFLATED INSIDE THE AIRCRAFT. This can not be emphasised to passengers enough. In the Ethiopian Airlines ditching, passengers inflating their lifejackets inside the aircraft was the probable cause of many deaths. If there is water inside the cabin, this prevents people from diving down to get out of the doors. They simply float upwards and get stuck on the ceiling. In aircraft with small exits, it will prevent people from getting through the exit. Anyone behind these passengers will also be trapped.

All luggage must be stowed safely. Think about evacuating if the aircraft is full of water and sinking. The last thing you want is luggage that has escaped from its stowages strewn all over the cabin. Why not, in large aircraft, put it in the toilets?

Think about the conditions you are about to be exposed to. Anything that you can take to aid the survival attempt may be a lifesaver. Encourage passengers to put on any warm clothing that they have with them. The more clothing they can wear, the better. Think about covering the hands and especially the head. The majority of body heat is lost through the head. Do not let anyone carry anything off the aircraft; this will hamper the evacuation process.

The most important items to take are the liferafts, the survival kits, including any signalling equipment, and the first aid kit. Anything else that could be useful may also be a bonus. Take water, if there is any at hand and if there is enough, encourage everyone to drink before the impact occurs. This will build up people's water supplies. Sea sickness pills are also vital. If they are accessible, hand them out before the impact. They take about 20 minutes to work and if you wait until in the raft before taking them, especially in rough seas, everyone will be vomiting before they have a chance to work.

Make sure all the essential items are stowed safely but very close to the door. You may need to retrieve them in the worst of circumstances and will have to be able to grab them quickly and throw them out. Equipment left behind is useless.

When it comes to the actual landing, make sure everyone knows the brace position and that their seatbelts are securely fitted. Make use of anything soft, for example, pillows and blankets, as added protection against impact injuries.

As crew, you are also responsible for your own safety. If you do not survive, the chances of your passengers surviving, without your guidance, are reduced.

This book does not go into brace positions in any detail because different seats within the aircraft require different brace positions and there has been no specific study of crew brace positions. It is generally suggested that cabin crew should be sat upright, with their heads towards the direction in which the aircraft is travelling. This means that if they are sat facing the back, their heads should be upright; if they are facing forwards, their heads should be tipped forward. By keeping the hands out of the way, for example, sitting on them, you reduce the danger of the arms flailing during impact.

It is important that passengers in fixed wing aircraft have their feet flat on the floor and slightly underneath them. This is to prevent the legs from flailing forwards during impact, one of the worst causes of serious injury which may prevent people from escaping the aircraft. In rotary wing aircraft, however, feet should be flat on the floor with the legs vertical. The feet should not be placed under the seat at all because in a helicopter, the impact force will be acting downwards rather than forwards and down as in a fixed wing. If the legs are back and under the seat, and the seat collapses, the legs will be trapped.

Sit level in your seat, with your bottom right into the seatback. Even a wallet in a rear trouser pocket will be enough to twist the pelvic girdle. In an impact, especially when the impact force is heavily downwards, the spine will not absorb the impact but will twist and could break.

Make sure seatbelts are fitted tightly and that the webbing is not twisted. Webbing is surprisingly sharp and in an impact will cut through the body like a knife through soft butter if an edge is left against the body. Make sure the buckle of the belt is easily accessible. On a four or five point harness, it should be in the middle anyway. On a two point harness, make sure it is near the middle. If it is too far round one side, you may not be able to reach it once an impact has damaged the seat around you.

The belt should be tight, especially the lap part. If it is too loose, it could twist over in an impact, putting the buckle facing the wrong way. If this happens, you will never release it. Also, a lap belt that is too loose could cause the person wearing it to slide under the belt in the impact, causing serious abdominal injuries.

Make sure your own lifejacket is on and fitted correctly. It should be a brightly coloured crew jacket, if carried on the aircraft, so passengers can identify you easily from a distance.

Pilots, and anyone else who is wearing them, should make sure headsets are removed before the impact. They will hit you in the face hard if they are not removed and if the aircraft ends up under water, they may get caught up and prevent the wearer from escaping.

Getting the Aircraft Down

Be ready for the evacuation. Underwater escape, where aircraft occupants have to fight their way out of a submerged aircraft cabin, is covered in the next chapter. Hopefully, if the landing was controlled and straight forward, the situation will not be as extreme as this.

Make sure you know where your exit is without looking. If the aircraft fills with water suddenly, you may have to open it with little visibility and submerged in very cold water. Run through your plan of action in the last few minutes before the impact.

The key is to get the evacuation underway immediately the aircraft comes to a halt. You do not know how long the aircraft will remain on the surface so get everyone out as quickly as possible. If there is a problem getting the rafts deployed, urge people to evacuate into the sea. Do not keep people waiting inside the cabin; they may be waiting to sink with the aircraft.

The evacuation is almost sure to be an 'each for his own' situation. As crew, all you can do is to remain calm, be assertive and get people off the aircraft quickly. You also need to get yourselves and as much equipment as you can, off the aircraft. Once everyone is in the water, or in rafts, you need to assert your knowledge of survival because this is where the real survival test begins. These issues are covered in greater depth in the next two chapters.

5 Underwater Escape

This chapter may seem appropriate only for helicopter pilots who are more than familiar with underwater escape and 'dunker' training but in reality, it could apply to anyone in any sort of aircraft, who is forced to make an emergency landing on water. Here, we take underwater escape literally. It applies to anyone who has to escape from an aircraft that is under water.

While helicopters frequently end up inverted in the water, necessitating the crew and passengers to escape from a water-filled cabin, the helicopter is usually kept at the surface by buoyancy packs stowed on the aircraft's floats. If the buoyancy packs inflate properly, the aircraft will not sink and this allows the occupants to escape from the cabin.

This is not always the case, however, and is never the case in fixed wing aircraft as they are not fitted with any kind of buoyancy aid. It is not uncommon for aircraft, especially general aviation aircraft, to 'nose in' on impact, either turning over or sinking. The force of the impact will, more often that not, smash the windshield and windows, thus allowing the cabin to fill with water. This then becomes an underwater escape situation although an added time factor is put on the occupants beyond holding your breath - the time in which it takes the aircraft to sink.

It is harder to form a picture of how airliners behave when landing on the water. From the experiences that we have seen, however, it is fair to say that occupants involved in a ditching face every chance of being in an underwater escape situation.

The Dutch Antillean DC-9 that ditched in 1970 remained largely intact and floated long enough for some passengers and crew to escape. Water did enter the cabin very quickly, however, and the entire aircraft had sunk out of sight within ten minutes. If the aircraft structure had been more damaged then water would have entered the cabin far more quickly. If there had still been anyone on the aircraft after ten minutes, they would have been dragged down under the water.

The Ethiopian Airlines Boeing 767 that was forced to ditch off the Comoros Islands in 1996, hit the water after a partially uncontrolled landing. The Captain was fighting with a hijacker at the time and so did not approach the landing in the controlled manner that he would otherwise have done. The aircraft wing hit the water first, spinning the aircraft round and the fuselage then broke up. Water flooded the cabin on impact and everyone who escaped did so from being submerged in water. Twenty six per cent of those who died of drowning on this aircraft had no other injuries. They simply failed to get out of the cabin.

Aircraft almost always sink eventually. The Royal Air Force (RAF) Nimrod that was ditched in a Loch in Scotland remained intact and floated long enough for all on-board to evacuate into rafts but did sink after a period of time.

The rule is to evacuate the aircraft, by any means, as quickly as possible because the aircraft could sink at any minute and with no warning. While, ideally, everyone wants to get into rafts without getting wet, the reality will probably be that escape through any exit straight into the sea will be the only thing that saves lives. If the cabin is submerged, then a quick escape by any method possible becomes the only life-saving option.

It is likely, if the aircraft cabin has filled with water, that the aircraft is not upright. It may have its nose tipped into the water, in which case it will be at 90 degrees to the sea. It may have inverted, in which case you will be upside down.

Training for underwater escape.
Here the simulated cabin is totally inverted.

The added problem here, beyond being submerged in water, is disorientation. No-one should underestimate the effects of this. It can make the difference between getting out of the aircraft and being trapped inside for ever. What effectively happens is that disorientation occurs, making it impossible for the person inside to find their way out. They may have absolutely no idea of where they are or even which way up the aircraft is. It can be so severe that it makes the advice in the next chapter, to ensure that you are clear of the aircraft before inflating your lifejacket, anything but obvious. You can think you are outside the cabin but really still be inside it.

Aside from impact injuries, disorientation is the reason for the majority of deaths inside submerged aircraft. This problem occurs in small aircraft and helicopters where often the person is sitting right next to their exit. Imagine the problems in a large aircraft, where people have to travel a distance to an exit. Obviously the chances of escape are severely reduced but the feat is not impossible.

The key to all this is planning, preparation and, most important of all, having a reference point as to where the nearest exit is. The other key is DO NOT PANIC. Panic will ensure failure.

It is vital to have a routine worked out in your mind as to how you are going to go about escaping before the impact occurs. Research shows that in helicopters, those who have undertaken dunker training have an 80% greater chance of escaping a submerged aircraft alive. This is simply because they have thought about what they would do and have run through the sequence in practice. They are prepared because they have considered the problem and have practical experience of overcoming it.

Keep calm and think. Know what your exact movements are going to be before the impact. Know where the exit is. This does not mean having a rough idea of where the door or hatch is, it means being able, if it is within reach, to find the exit handle and open the door with your eyes closed, from your seat, very quickly. This is vital because in a minute, you are going to have to do it in the dark and probably upside down - and your life is going to depend on it!

The first thing to be sure of is that you are sitting properly and that your seatbelt is securely and safely fastened. Follow the advice given in Chapter 4 concerning preparing for the impact.

Know which way you have to twist or lift the buckle of your seatbelt, and which way to pull it free. Do not assume. Belts do differ and you will only have a split second to release it.

If you think this is obvious, do not under-estimate the shock of the situation you will be in. One of the most common scenarios (observed by the author several times simply in training) is for people to try and evacuate under water without undoing the seatbelt at all. They make all the right motions of getting out but forget to unstrap themselves. They fight against the obstruction without realising what it is. It has been a cause of people drowning, trapped inside sinking aircraft.

Never inflate your lifejacket inside the aircraft. The plan of action should be to brace, knowing that you can find the seatbelt buckle with one hand and the exit handle with the other. This should have been practised in the minutes before the impact. Wait for the aircraft to come to a halt before you do anything. If the aircraft is filling with water, the pressure of this water will be severe. As soon as things begin to settle down and the bubbles stop shooting past your nose, this is the time to act.

Find the exit handle and once you have got hold of it, do not let go. Undo the seatbelt with the other hand and release the exit handle. The door should open freely because of the water pressure from inside the cabin. If it does not, because the pressure on the outside is greater than that on the inside, you will have to push the door. Maintain your reference points at all times and never let go. If possible, put your foot on something solid that you can use for purchase to push against. As the door opens, pull yourself out. Never let go of the exit because you may not find your references again.

Do not swim inside the aircraft; rather pull yourself along. If you are wearing anything that makes you even slightly more buoyant, for example, air trapped inside clothes, you will find yourself floating to the surface and will find it hard to pull yourself down again. Also, swimming will use up energy that will prevent you from holding your breath for as long.

Once you are free of the aircraft, put your hand above your head as you float to the surface. That way, if there is anything above you, for example, aircraft parts or people, your hand will hit them first, not your head.

Dunker training courses are the perfect way of preparing for underwater escape situations. Such courses allow you to experience the reality of the situation while helping you to prepare fully for the real thing. Although compulsory for many helicopter crews, for example, those working off-shore, these courses are also highly recommended for General Aviation pilots. Airline personnel, also, would benefit greatly from such an experience, especially Safety & Emergency Procedures (SEP) instructors.

6   In the Water

The ideal, in a ditching situation, will be to step from the aircraft into the liferaft but in reality, this will not always be possible. The perfect situation will be for the aircraft to float long enough that you can get everyone into the slide rafts or get liferafts deployed alongside the wing and for everyone on board to get into a raft without getting wet.

It is very dangerous, however, to plan a ditching evacuation with the assumption that you must not get wet. By doing this you may put yourself at risk in your determination not to get into the water.

Damage to the aircraft will almost certainly occur and this will cause it to sink; if not immediately then eventually. As has already been seen, occupants of the aircraft should therefore get well clear of the fuselage as quickly as possible and this may well involve swimming.

There are those who say that the aircraft is the best shelter - stay with it for as long as possible, but the aircraft could sink at any minute; very quickly and with very little warning. Anyone left inside will be trapped and will go down with it.

The deciding factor is how long it takes to get everyone out. You can not allow the situation to occur whereby passengers are waiting inside the cabin for you to deploy a raft. If you do not have a normal evacuation situation, where passengers are evacuated quickly into slide rafts, or they can not exit immediately into a liferaft, then it is actually safer for them to get into the water.

In smaller aircraft, where there are no slide rafts, the same rule applies unless the aircraft is very stable and it is possible to deploy the raft from the wing. If you are trying to do this, though, make sure everyone is waiting outside the cabin at all times.

Apart from the risk of people being trapped inside the sinking aircraft while you are struggling with a raft, the other danger of trying to deploy liferafts from the wing is that there are many sharp objects on the wing and around the fuselage that can cause damage to the raft. Once a raft is punctured it is useless. Even those rafts that have two inflatable chambers will offer very little protection once a chamber is totally lost.

Getting into the water, then, is technically the last thing you want to do; in the water you are in danger of drowning and of contracting hypothermia, both of which will kill in minutes. In practise, however, you are very likely to end up in the water, even if it is just for a short time. As long as exposure to the water is short and the appropriate actions are taken once the raft is reached, this need not be a life threatening action.

Studies of actual incidents have shown that the vast majority of survivors have ended up in the water before getting to their liferaft. Swimming may risk your life - but it may also save it by preventing you from being trapped inside a sinking aircraft.

If avoiding the water is not possible, then the next best advice is staying in the water for as short a time as possible - and the next best is knowing how to look after yourself while you are in there.

The quickest killers are drowning and hypothermia and any person in the sea is vulnerable to both. If you have to get into the water, get out of it again as soon as possible. The thermal conductivity of the body is 26 times what it would be out of the water so it is imperative, to preserve body heat, that the least time possible is spent submerged. Children and babies suffer far more quickly than adults; they have such a huge body area to weight ratio and so they lose heat very quickly. Because they are so small, they do not have a reserve.

Although putting them into a child or infant lifejacket, this will not help maintain their core temperature. The infant life preservers that are now on the market; an orange inflatable and sealed tube in which babies are totally protected from the elements and from drowning, are an excellent idea for young children. Hoover Industries is the current largest manufacturer of these infant life preservers. Older children should be wrapped up in a lot of clothing and kept out of the water as much as possible.

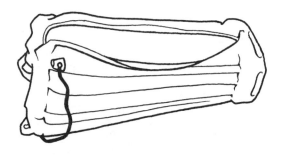

Infant life preserver.

To give an idea of survival times in open water, if a person is young and healthy, they can expect to survive up to nine hours with a lifejacket at 17 degrees C. This is the UK's warmest summer sea temperatures. In colder water, for example, six degrees C, you will only last between 45 minutes and two hours with a lifejacket. As the water temperature decreases, the survival time will decrease proportionally. Without a lifejacket you will not last more than nine minutes in 10 degrees C - and water gets much colder than this!

It is important to relate this to likely rescue times. It is generally said that rescue will come up to three days after the ditching. This depends on where it is in the world that you ditch and on how well the Search & Rescue system that is looking for you is organised. Unfortunately, the system works such that those who are best equipped for survival, such as the military, will be picked up the quickest because their aircraft tracking is so good. Those who are least prepared, such as the airlines, will be picked up the slowest, because, despite air traffic control, they are not nearly so well tracked.

If you ditched a light aircraft en-route across the English Channel, for example, you could expect to be picked up well within 24 hours whereas if you ditched a Boeing 747 mid-Atlantic or mid-Pacific, you are likely to have to wait for the full three days.

Here is an example. In January 1992, a two-man crew ditched a Beech Bonanza in the North Sea, not too far from land. This crew were able to relate their position before ditching very accurately from an on-board GPS. They were spotted by a C-130 Hercules aircraft and were picked up by a nearby fishing trawler that was sent by the aircraft to help. Both crewmembers were wearing immersion suits and lifejackets. Both had been in the water before climbing into their liferaft. Both were picked up with hypothermia. Despite the GPS, the fact that they were close to land, the Hercules crew seeing them quickly and the closeness of the fishing trawler, their rescue still took eight hours!

That statistic of lasting a maximum of nine hours in warm water without a raft suddenly does not seem so good!

Never under-estimate rescue times. Chapter 10 deals with search and rescue from the water in detail but it is important to realise that rescue usually takes far longer than most people ever dare imagine. For a rescue team, knowing where you are and actually finding you are two totally different things. Rescue services may know you have gone down in the middle of the Pacific; a long-range search and rescue aircraft may come out to you and may circle overhead dropping you supplies but you will not be rescued until a ship, with helicopters on-board, gets to you up to three days later. Even then, actually finding you is another story. Seeing a liferaft from the air, let alone a body bobbing around on the water's surface, is not at all easy.

DO NOT BE COMPLACENT. So many people are complacent about survival and rescue and this is very often fatal. It is your duty to do everything you can to protect yourself and your passengers because rescue may not come as quickly as you first hope.

Getting into the Water

When it becomes evident that you will have to get into the sea, inflate your lifejacket, if you are completely clear of the aircraft, before going in. It may be that you were submerged inside the cabin, in which case you should make totally sure you have escaped the aircraft before pulling the inflation toggle. Remember, it is easy to become disorientated under water and you may think you have exited the aircraft when in fact you are still inside it.

Unless you are forced into the water suddenly, try to enter the sea slowly. This allows the body to adjust to the shock of the cold more gradually which may even prevent heart attacks in the vulnerable. Cover your nose and mouth as you enter the water because it is easy to gasp on contact with the cold, forcing water to enter the nose and throat. This can easily lead to drowning or to a spasm in the throat that causes the epiglottis to close, thus not letting air into the lungs.

Secondary drowning can also occur, which is caused by a swelling of the lungs from the irritation of water entering them in the first place. This can happen up to ten days after the event, even though the casualty can make an apparently complete recovery at the time.

As well as covering the nose and mouth, it helps to consciously hold your breath when entering the water but do not take too big a breath beforehand. A normal breath will suffice. If you fill the lungs completely, the pressure on the diaphragm stretches it and makes it want to contract, urging you to breath out and in again more quickly.

Once in the water, this state should be maintained, with the nose and mouth still covered, until you have stopped bobbing up and down.

If it becomes necessary, for any reason, to enter the water from a height, make sure first that your entry is clear; that your inflated lifejacket is firmly held down and that your feet are together on entry. The maximum safe height for jumping with an inflated lifejacket is six metres. Any higher than this and you are in danger of breaking your neck upon entry from the lifejacket forcing its way upwards. If, then, as an example, you are forced to jump from a Boeing 747 upper deck door and the slide raft is not inflated or has been detached from the aircraft, jump before inflating your lifejacket.

Once in the water, swim away from the aircraft, the debris and the other people who are trying to get into the water from the aircraft. Swim on your back and use only your arms. Be prepared for fuel on the water; swim away from it and avoid burning fuel at all costs.

In the unlikely event that you are trapped by burning fuel you may have to jettison your lifejacket in order to swim under the fire, which will only burn on the surface of the water, to escape. This is why lifejackets should be tied in a double bow and not a knot; so that they can be untied in a hurry.

The other situation in which you may need to jettison your lifejacket is if you are trapped inside a sinking aircraft. If the lifejacket were to inflate by mistake, by the inflation toggle catching on something for example, the only chance you have of escaping the sinking cabin is to get out of your lifejacket.

This may sound extreme but there were many cases during World War II where crews faced this exact situation and the only thing that saved them was diving down under the flames. Obviously, you only want to jettison your lifejacket as an absolute last resort because without it you will severely cut down your survival time.

A situation in which everyone has to take to the water is going to be very much an 'everyone for themselves' situation. There is little, as crew, that you can do in this case except urge passengers to leave the aircraft, inflate their jackets and get into the water. In reality, you may find that the situation is so confused that you don't even see any one else while the evacuation is taking place.

If there is no raft, then the swimmers must remain in the water until rescue comes. 'Swimmers' is a loose term because it is very important that you do not try and move in this situation. Every surface area of the body loses heat to the surrounding elements; the greater the surface area exposed to the water, then, the more heat is lost.

By closing up as small as possible, you will preserve valuable warmth. Everyone should keep their legs crossed from the groin - this is an area from where a great deal of body heat can be maintained if the legs are closed and lost if the legs are opened.

Try this in a swimming pool. Cross your legs from the groin and keep still for a while. Once you begin to feel a bit cold, open your legs and you will feel the rush of cold as the trapped body heat escapes.

Once in the water, it is the crew's job to try and get everyone together in one or more groups and, if possible, into the rafts as quickly as possible. Keeping together is vital for several reasons. These include the following

*   Morale is greatly boosted by people keeping together.
*   Strong swimmers can aid non-swimmers and the injured.
*   The more people who huddle together, the more body warmth is shared.
*   The bigger the group of people in the water, the easier it will be for rescuers to see them.

Conserving body heat in the water.
Make sure legs are kept tightly together or crossed and
the elbows are down by the sides, protecting the groin and armpits.

Huddle together in the water with injured parties, children and non-swimmers in the middle. Although you do not need to be able to swim in this situation - you have a lifejacket on and do not want to move anyway - being in the middle of the group will give the non-swimmers confidence. They are less likely to panic and thrash about or try to grab hold of other swimmers, risking life.

Arms should be linked with each other and held with the hands at shoulder height so that the elbows can be kept in close to the body. This will ensure that heat is trapped into the arm pits. This is illustrated in the picture on the next page.

**A group huddling together to share body heat. Note that their arms
are linked, with their elbows closely in to their sides.**

Entry into cold water will cause the heart rate and blood pressure to increase
and the surface blood vessels will shut off automatically, keeping the blood
around the essential organs of the body. This is the body's natural defence to
preserve heat. After about half an hour, however, this protective reflex will fail
and the blood vessels at the extremities will open up again, allowing the
remaining heat to literally flow from the body. If you have liferafts, then, you need
to try and get everyone to them as soon as possible.

When swimming to get to the rafts, or away from danger, your group should
form a line - with the strongest swimmer at the head of the group. The whole
group should swim on their back, using only their arms and keeping their legs
wrapped around the waist of the person in front. Keep injured people in the line
and make sure they are aided, where possible, by strong swimmers.

Using Seat Cushions for Flotation

As a last resort, seat cushions can be used for flotation in those aircraft where
they are detachable from the seats for this purpose. The two straps on the back
are designed for holding on to but it is important that when you put your arms
through the straps, you do not hold on to the straps themselves, but link your
arms and hold onto your wrists. This is to ensure that you do not involuntarily let
go of the cushion through cold.

Flotation with a seat cushion.

It is also possible to group together using seat cushions. The same principles should apply. Link arms and keep the elbows down to preserve body heat from the armpits. If two people are linking together, the cushions should be held behind the other person, with your arms round them, to ensure that the bodies of the two people are together to share heat, rather than having the cold cushions between you. It is also possible to put a third, injured party or non swimmer, between the two people. Try to avoid fuel spills in the water as seat cushions are easily corroded by fuel.

In desperation, the old trick of removing trousers, tying knots in the ankles, blowing them up from the waist and using them as a flotation aid, may help keep you afloat for a short period of time.

Improvise a Raft

It is always possible to improvise a raft out of debris. Although this is not going to hold many people, it could be used by small groups to get them out of the water for a short period of time if there is nothing else available.

First Aid in the Water

There is little you can do in the way of administering first aid to injured parties while in the water. As a crewmember you are there to help and to control but you must also be aware of preserving your own life. Without the crew, the passengers will have a far smaller of a chance of surviving because they are relying on you to help them by putting into practice your SEP training.

Passengers with broken limbs can be helped to get to the rafts and to board them; bleeding wounds can be held and elevated; unconscious bodies can be towed to the rafts and even be resuscitated on the way but it is important to assess how much you can safely do without jeopardising your own and the rest of the group's safety. Try resuscitating someone while swimming in a swimming pool and then imagine trying to do it in a rough sea! You can concentrate on first aid once you are in the raft. The important thing at this stage is to make sure you are alive to get there.

Equipment

Clothing can make all the difference between life and death if it becomes necessary to get into the water. Even if you avoid this option, and manage to get straight into the liferaft, the warmer the clothing you have on, the better your chances of avoiding exposure and hypothermia.

The criteria are different depending on the type of aircraft and the reason for the flight. Those, for example, who are working in the off-shore industry, including the military, will automatically wear immersion suits and lifejackets at all times while flying over water. These people will be best prepared for a ditching.

Private pilots are also encouraged to wear such clothing, or at least have it with them in the aircraft. Many do not, however, as they do not consider their short pleasure flight over water to be in a high risk category. It has to be pointed out that time after time, incidents prove these pilots to be wrong. In fact, the safety statistics of countries throughout the world are littered with cases of General Aviation aircraft ditchings.

As a rule, those flying in rotary-wing aircraft should wear their emergency clothing at all times, where possible, as there is a greater statistical risk of accidents in rotary-wing aircraft than in fixed wing. In addition, as a general rule, there is usually less time to prepare for such accidents in helicopters than in aeroplanes. Sometimes ditchings can be very sudden.

Commercial aviation crew and passengers, as usual, are the least prepared. Airline aircrew will never have access to an immersion suit. The risk of ditching is considered so low in commercial passenger aircraft that many 'extra' safety items are just considered unnecessary. The cost of buying them in the first place, and the cost of the extra weight of having them on-board, render these items non-cost efficient.

It has to be remembered, though, that however low the risk, it is still there and if something does go wrong, the airlines, responsible for the greatest number of passengers, will be the least prepared aviators to deal with it.

Even without the ideal equipment, crew and passengers can protect themselves by wearing as much extra clothing as possible. Donning this clothing should be one of the items in the 'preparation' phase; talked about in Chapter 4. Any clothing that may be on-board will help although natural fibres are generally better than man-made. Wool is the most protective fabric in the water; it retains 50% of its insulating properties once wet, compared to cotton which retains only 10%.

The head is one of the most important parts of the body to cover, as a huge amount of body heat escapes here. Layers are also important, as each will trap warm air, every layer acting to insulate the body.

Cabin crew uniforms are hardly the most suitable for a survival situation of any kind. Those made of wool are best as even though the designs are usually unsuitable, the fabric will at least provide some insulation. Overcoats, hats, gloves and anything else that can be found should be worn on top of uniforms. Crew lifejackets, in the majority of airlines, are a different colour to passenger ones so you do not need to worry about keeping the uniform recognisable to passengers.

Life Jackets

There must be one of these for every aircraft occupant, regardless of the type of aircraft or its mission.

Lifejackets must NEVER be inflated inside an aircraft and this MUST be impressed upon passengers in safety briefings and again, if at all possible, in your 'preparation' phase. Inflated lifejackets will prevent you from escaping a submerged or semi-submerged aircraft. You will float to the 'ceiling' of the cabin, which may be the floor, wall or ceiling, depending on the orientation of the aircraft at that point. IT WILL BE IMPOSSIBLE FOR YOU TO DIVE DOWN UNDER THE WATER TO ESCAPE THROUGH THE DOORS. An inflated lifejacket will also prevent you from escaping through smaller exits, even if not submerged.

Not only will an inflated lifejacket prevent the wearer from escaping, it will also prevent everyone behind them from escaping because the person in front will be floating around blocking the exits. It is imperative then, that as crew, you do everything in your power to instil into your passengers that lifejackets must not be inflated inside the cabin. Do not rely on the pre-flight safety briefing to do this. Very few passengers listen to this briefing and they certainly will not remember that small fact among all the others in a panic.

Even in a larger, wide-bodied aircraft, where there are no small 'plug' type exits, lifejackets should not be inflated inside the aircraft. This book has already discussed the chances of the aircraft sinking. Larger doors will make no difference to escaping if you have to dive down to find the exits. Once water has entered the cabin, more will follow and you will soon find yourself unable to get out.

It is important that lifejackets are fitted correctly and again, passengers should, if possible, be helped with this. Improperly fitted lifejackets can result in the jacket popping up over the head once in the water, or can cause problems when the person is being winched by a helicopter. As a general rule, they should be tied as tightly as possible to prevent them from rising up, forcing the occupant down.

Airline lifejackets are very different from General Aviation jackets or those used by the off-shore industries or the military. The airlines require the most light-weight jackets possible and, of course, it is hoped that they will spend their entire operational life rolled up in a pouch under each seat. They are not strong and they are not durable - in essence, they are designed cheaply for one-off use.

They come in two styles; the double chamber which is favoured in the United States and the single chamber which is favoured in the United Kingdom. Both have benefits and disadvantages; with the double chamber the wearer has another chance if one of the chambers is pierced. With the single cell vest, if it is pierced there is nothing else to rely on. The double chamber jacket, however, has a gully down the front which channels water and fuel straight down into the swimmer's face. The single chamber vest, however, deflects this water away from the face. Each jacket should have a light, a whistle and a manual blow-up tube for each chamber.

For General Aviation (GA) pilots and passengers, however, these lifejackets are not suitable as they are not durable enough for continuous wear. It is important for crews and passengers on such aircraft that they choose their lifejacket carefully. They should choose a proper lifejacket, which will keep an unconscious body on their back, rather than a buoyancy aid which will not and which often has a permanent buoyancy. They should look for aviation jackets and not the 'automatic inflation' type which inflates upon contact with water that is often found in the marine industry.

In addition to the basic design of the lifejacket, whether airline, military or GA, a spray hood is a life-saver. A lifejacket will keep you the right way up in the water, even if you are unconscious, but it will not stop waves from breaking over your face. This can easily cause drowning, especially in the unconscious or once fatigue has set in and resistance to danger is lower. Spray hoods will prevent this.

Almost all jackets used by the military have spray hoods; General Aviation pilots should choose one of the many makes of lifejacket that do feature them and the airlines, once again disadvantaged, are very unlikely ever to see one. The benefits of this simple device should be borne in mind, though, because one of the two greatest dangers - drowning - is significantly reduced if you have a spray hood on your lifejacket.

Land/Shark Survival bags

These bags are made of aluminised film reinforced ripstop fabric that has been laminated and are coloured the International Safety Orange. They are designed to reduce the loss of vital body heat, lessening the risk of hypothermia. They are suitable for use in the water, in a raft or on land. The aluminium coating reflects up to 80% of radiated body heat and the bag is designed to enclose the swimmer right over the head, including lifejacket, without restricting movement.

The bag also serves as a deterrent against shark attacks as it prevents body wastes and fluids from escaping into the surrounding sea; something that could attract sharks from several miles. It also changes the profile of the body away from that of an injured creature.

There is a tether attached to the bag, which helps people in the water to stick together. This bag is designed to increase survival times by reducing heat loss. The author can certainly vouch for it reducing the loss of vital body heat.

Again, try this in a swimming pool. Get into one of these bags, fasten it over your head and float as still as possible in the survival position. Keep still for some time and you will probably begin to feel a bit cold. Then remove the bag and as the body heat rushes out and the cold water rushes in, you will realise just how warm it has kept you.

The only downside to these bags is the disorientating effect they have on the person inside. This is due partly to the fact that for them to work most effectively, they should be closed up over the head, preventing the occupant from seeing anything, and partly due to the waves trying to roll the person in the bag over.

7   Liferaft Management

If you are fortunate enough to be in the rare situation whereby you can get yourself and everybody else into your liferaft directly from the aircraft, you are in a good position. The most important danger to be aware of is damaging the raft and so take precautions against this at all costs.

The raft needs to be positioned as close to the exits as possible so it can be obtained easily in an emergency. There have been so many cases of people getting out of aircraft but leaving their rafts in stowage compartments or in the back of the aircraft. They are useless here. Rafts, along with survival equipment must be just a hand-hold away or they are in danger of being left, especially if the aircraft breaks up and fills with water. EQUIPMENT LEFT BEHIND IS USELESS!

Do everything you can, without jeopardising your own safety, to get your raft out. Even if you are using slide rafts, liferafts should be taken with you if you have them on-board because they will offer so much more protection than sliderafts can. Designs of raft will be discussed further in Chapter 8.

Deploying a Raft

Throw your raft into the water as far away from the aircraft as possible. Make sure it is attached to something, even if this is just yourself, because if it floats off, which it will do very quickly in the wind and current, even the strongest swimmer will never be able to catch it up.

You will need to pull out a surprisingly long amount of line before it inflates. Do not be disheartened if the raft seems to be taking a long time to start inflation; you have just not pulled out enough line. Remember that the raft does not have to be fully inflated to allow boarding. After about 17 seconds, there will be enough air in the raft to allow someone to board it safely.

Boarding a Raft

When boarding, whether from the aircraft or the water, it is essential that you keep the raft balanced and this means positioning all occupants evenly around the raft to help prevent capsize. If entering from the water, the strongest people should board first and should position themselves either side of the entrance to help others to board. These subsequent occupants should then position themselves at the back and then sides of the raft - one to the right and one to the left - to keep the weight balanced all round.

**Boarding a raft is not as easy as it looks,
as this training session shows.**

Boarding a raft from the water, especially with an inflated lifejacket on, is not easy. Almost all rafts are designed with an overload limit that allows them to carry in excess of their recommended occupant limit. This overload usually caters for double the recommended number of people. Whatever the situation, no-one should be turned away.

Swimmers in the water who have made it to the raft and who are waiting to board should make use of the lifeline that runs around the raft. You should not hold onto it with your hand because in the cold you will be likely to let go and not be able to take hold again. Instead, you should link an arm through the line and await your turn to board. It is common for people to drown because they can no longer hold on to something due to body malfunction because of the cold.

Should the raft have inflated upside down, it is obviously essential to right it before boarding can begin. This involves one or more people, depending on the size of the raft, taking the boarding line, which runs across the bottom of the raft, standing on the edge of the tube and pulling the raft over towards them. Always try to do this with the wind's assistance. If you try to pull the raft against the wind you will get no-where. This is an exercise that needs practise and it may need more people than you think to pull the raft over.

Raft Management

Once in the raft, discipline is essential and one person, a crewmember, should take control. Ideally you should try and get at least one crewmember in each raft. It is usually after the initial incident, when people have got to relative safety but are still vulnerable, that the truth about their situation dawns and then panic and confusion can set in. This can be fatal in itself so it is important that everything is done to try and avoid this. This is where survival routine comes in. Look at this as the next phase of your journey.

Good management within the raft is vital. You are not safe once you are in a raft, you are just safer. You still have a long way to go. REMEMBER: YOU ARE NOT A SURVIVOR UNTIL YOU HAVE BEEN RESCUED and A RAFT IS ONLY AS GOOD AS ITS OCCUPANTS.

There are certain tasks that must be carried out. Four terms can be used that encompass these: Cut - Stream - Close - Maintain.

Cut: Cut the painter that is securing your raft to the aircraft as soon as possible. Hopefully everybody will be aboard the raft by now but the first people in the raft must be prepared to do this as soon as the aircraft shows signs of sinking, whether everyone is on-board or not. Otherwise the raft will just be dragged down with the fuselage. It is important that someone is always keeping an eye out for this.

Once the painter is cut, the raft should be allowed to move away from the immediate vicinity of the aircraft. This will happen automatically if there is a wind or strong current; if there is not, the raft paddles can be used to manoeuvre the raft away. This will help protect the raft from damage from the debris and fuel in the water.

Stream: The drogue or sea anchor should be deployed as quickly as possible. This will slow drift and will also help to prevent capsize. If there are still swimmers in the water, putting the drogue out as soon as possible will slow the raft to enable these people to be picked up. Although the raft is at the mercy of the winds and currents, the closer it can stay to the downed aircraft, the easier it will be for rescue parties to find it. They will usually start looking close to the wreckage and then will calculate the raft's movement from there according to the winds and currents. The closer you are, the shorter your 'survival' ordeal will last.

The drogue will also twist the raft round so that the entrances are not in the line of the wind, thus increasing the protection of the occupants.

Close: The raft entrances should be closed as soon as these other functions have been carried out and it is certain there are no more swimmers in the water. In the case of the raft with the additional, tent-like canopy, this should be erected now.

All the time the entrances to the raft are open, heat is escaping and wind is blowing in to further cool down the occupants. As soon as the raft is closed up, the occupants can begin to share body heat and the inside of the raft will begin to warm up. It is very important, however, to allow the carbon dioxide, which will have built up in the raft, to escape every half hour. If there is a look-out point on the raft canopy, a tube that allows the person on look-out duty to see out but ties up closed when not in use, use this to let the carbon dioxide escape. Otherwise, simply open the entrance, just a fraction, for a couple of minutes.

Maintain: This involves all those other tasks that must be carried out in a raft for the duration of the time before rescue arrives.

There is a very strong psychological point to this as well as a practical one. The most important aspect to survival is the will to live and the determination to survive. Keeping everyone busy will help morale enormously. Not only will everyone's mind be taken off their plight but they will also feel that they are contributing to the success of the survival mission. Once people stop they often give up. In a life and death situation, once they give up, they often die.

To give an example, a World War II vessel was sunk and a group of survivors made it to the liferaft. This group comprised three officers and the rest men. The policy during war-time was for pay to be stopped once ships were lost as the men were no longer working. All the men in the liferaft thus refused to carry out any tasks in the raft because they were not being paid for it. The Officers therefore had to carry out all the raft management tasks until rescue came. The rations were shared out equally between all on-board during this time. Once rescue came, only the Officers were alive. All the men had died because they gave up and were not involved in their own salvation.

It is not over-reacting to remove all knives and anything that could be used as a weapon. You can never be sure what a desperate person will do. Knives can be a crucial tool within the raft however, especially for repairs so keep one or two but make sure they are very well stowed and watch any passengers who may be a threat either to themselves or to others.

The tasks that need to be carried out have to be prioritised according to the individual circumstances. An emergency action card is usually kept in the survival pack and can be referred to on entry into the raft.

First, where there is more than one raft, the rafts should be tied together. Use the raft painters to do this. Tie the rafts about 10 - 15 metres apart, to allow one raft to crest a wave while another is in a dip, without one capsizing the other. By tying the rafts together you are increasing your chances of being seen by rescue craft. In a choppy sea, at least one raft will always be in sight on the crest of a wave and the rafts will spread out to make a line. This will help them stand out from the air. You are also increasing morale by keeping everyone together.

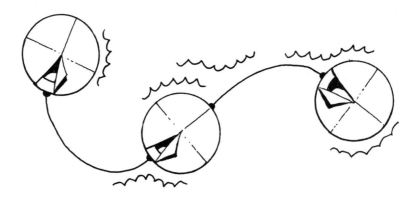

Rafts tied in a line.

Protection from Cold

The basic rule of Protection and Location as the first two most important things should be observed. The raft is free from the aircraft and the drogue has been set. The raft entrances are closed. This is the first step. Next, the occupants must be further protected from the effects of the cold and the wet.

As for the cold, inflate the raft floor, if the floor is inflatable. This will provide a good insulating layer between you and the water. Bail out the water that is in the raft and sponge the floor dry as much as possible.

Put on any extra clothing that can be found. Some survival kits have one or two survival suits or blankets that can be used to protect the most vulnerable; the young, old or injured. Huddle together and share warmth. Macabre as it sounds, clothing should also be taken from the dead; they have no need of it now and it may save you from going the same way. Remove any dead from the raft - they are taking up valuable space and will not improve the morale of the survivors.

Simple exercises, such as wriggling the toes and fingers will help improve circulation without losing valuable body heat or energy. Keep the trunk of the body still to preserve warmth unless engaged in tasks for the good of the raft and its occupants. Remember, excessive movement will use up energy that you cannot replace.

Some people advise taking off all wet clothing but this very much depends on what the clothing is made of. Water in certain fabrics, especially those made of natural fibres, can actually warm up once out of the wind and can be insulating.

Keeping a Look-out

Another task that must be carried out, right from boarding the raft and until rescue arrives, is that of look-out. This is very important for many reasons: to warn of dangers, to look out for other survivors still in the water and to watch for signs of rescue.

Sea-sickness pills

Other tasks that should be undertaken include handing out sea-sickness pills if you did not have the opportunity to do this before ditching. It won't take long for occupants of a raft in a choppy sea to be vomiting and this loses water from the body. Sea-sickness pills usually take 20 minutes to start working so if they weren't taken before ditching, make sure they are given out as soon as possible.

Do not let people vomit in the raft; it will make everyone else start vomiting. Neither should people vomit over the side; this will attract sharks from miles away. The ideal is for people to vomit into bags. These can then be sealed and used as hot water bottles - a very effective way of warming the core temperature of the body!

Raft Repairs

Have someone ready to make any repairs to the raft and someone to sort out the kit and rations into organised groups before stowing these. Make sure everything that can be tied down is tied down in case of capsize or a large wave. Losing valuable equipment is not a great morale booster and can make the difference between surviving and not surviving.

First Aid

First aid treatment can also now begin. It may seem that this should be given first priority and indeed, every situation must be judged on the individual factors effecting it but remember one basic rule. The lives of the majority are more important than the few. It is not worth risking other passengers' lives by neglecting basic important survival tasks in the attempt to save one or two lives. The survival of crewmembers is also paramount. Without you for guidance, the passengers will have a much reduced chance of survival on their own. First Aid is covered in greater depth in Chapter 9.

Ensure that everyone has a task and that they know the importance of this task. Arrange things on a roster so that everyone has the chance to rest and to work.

It is advisable for everyone to urinate within the first two hours of being in the raft. Although you are trying to preserve fluids, it is important to minimise the risk of water retention which can lead to a number of medical problems.

Hot Climates

Depending on where you ditch, you may be struggling with the heat rather than the cold and this, too, can be debilitating. Heat exhaustion and heat stroke can be deadly. To keep cool, deflate the floor of the raft so the canvas is directly against the cool water, wet the sides of the raft and canopy and open it up as much as possible to let any wind through. If the sea is calm, pull up the drogue. This will line up the raft's entrances with the wind to increase wind flow through the raft. Take off clothes and wet them or even just pour cool water over yourself on a regular basis. Swimming should be avoided. The exertion will deplete essential body fluids and depending on where you are, there may be a risk of sharks.

Make sure that you protect against the sun's rays on the body by keeping covered with lightweight clothing. Sun cream is actually one of the most useful items to include in a survival or first aid kit because the sun can cause serious burns which, as well as being painful, will dehydrate the body very quickly.

Rations

One of the major factors in a survival situation will be the acquisition and distribution of rations. Although this is, in reality, a factor of little importance compared with getting out of the water, keeping warm and dry and being rescued, it will soon rear its head because food and water are a great morale booster and lack of them frightens people significantly which does not help the survival effort.

Water should be supplied in survival packs. Additional water may also have been brought from the aircraft. This should be stowed extremely carefully as loss is careless and critical. Water acquisition methods such as reverse-osmosis pumps will also be present in the raft and should be used. This equipment will be covered in more detail in Chapter 8.

You will be surprised to learn that water is not that critical unless you are in very hot climates. Think about it logically. If you are adrift for three days, which hopefully will be the longest that it will take for rescue to find you, the average human can last this long without water. 24 hours, then, should not be a problem.

This is obviously not ideal but in any case, water should not be handed out for the first 24 hours, except to those who have lost large amounts of body fluids. Those who are badly burned, have vomited violently or have been bleeding will fall into this category. In addition, the elderly and the very young will be at risk of dehydration first.

Following this first 24 hours, only half a litre of water should be distributed to each person every day. If there is not enough for this, then what water there is should be divided out equally during the next few days. The daily ration of water should be taken three times a day and should be swilled around the mouth before swallowing as the mouth will have become very dry.

Never be tempted to drink sea water. It may give short term relief but it will just make a person doubly thirsty very quickly. More seriously, the kidneys will be unable to cope with the salt content and will fail very quickly.

Do not drink urine, either. The body has spent the last eight hours getting rid of poisonous toxins and if you drink it you will replace them in five seconds. You can, however, distil urine or pump it through a reverse osmosis pump. This is not so much an issue at sea, where water is plentiful - albeit sea water - but is far more of an issue in a land survival situation. Water procurement of various sorts will be covered in the Desert survival section.

Food is even less critical and although supplied in the form of oatmeal biscuit and white glucose tablets, should not be taken unless there is a sufficient supply of water, ideally one and a half litres of water per person per day. All foods require stomach juices to digest them and this will increase the incidence of dehydration unless this fluid is replaced.

8   Liferaft Design and Equipment

A liferaft is going to be your best chance of survival in any sea survival situation. People have lived for months in rafts, using equipment that they both bought for the job and improvised along the way.

The design and quality of your raft and equipment is fundamental to the success of a survival situation, though, and it is a sad fact that money influences most buying decisions - not need. It is easy, when standing safely on dry land, to make these decisions without having any idea of the conditions you may be facing if you are ever in the unfortunate position of having to use it for real.

The idea of this chapter is to outline the differences in raft design; the pros and cons of various features and shapes, and to give an idea of the equipment that is available for use in the raft.

Shape

The shape of the raft is vitally important as it is this shape that determines its vulnerability to capsize. It is important that rafts do not offer any edge to the waves or wind that could flip them over. Rafts are like boats without power. Left to float in the sea they will lay 'beam-on' to the waves and are therefore more vulnerable to being flipped over.

Round rafts are always the best as which ever way they lay to the wind, they are never beam-on. The more oblong the raft, the greater the surface the wind and waves have to catch it and capsize it.

Sliderafts, then, are a disaster as they will lay continually beam-on to the wind and are very unstable. This is why liferafts should always be carried in addition to slide rafts.

Round rafts are obviously not totally round, but are made of many short sections all attached in a decahedron shape, for example, depending on their size. To give these rafts maximum protection against wind and waves, and to make the raft strong, each of these sections should be as short as possible.

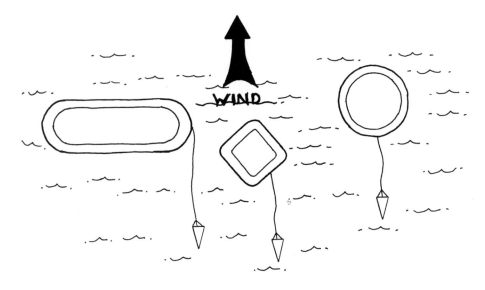

Different shape rafts in the wind.

FEATURES

The Canopy

Having established the stability of the raft, which is vital unless survivors wish to be thrown into the water repeatedly and lose their lifesaving equipment in heavy seas, the most important feature on any raft is the canopy.

A canopy is the number one lifesaver. Having escaped from the freezing water, protection from the wind is then essential. The Royal Air Force, for example, maintain that getting wet in the water is actually not as serious as being unprotected once you are out of the water. Water, once warmed, can actually act as insulation if the right clothing and fabrics are worn, whereas wind will cause a person to suffer from exposure in seconds.

The philosophy of many people is that they buy the raft, because they feel they need one, but they cannot afford a canopy so they do not buy this. This is a total waste of time unless you are only ever flying over warm climates where the seas are constantly very warm and not rough.

There was an incident in 1982 where five people took to a dinghy because their yacht sank. The dinghy had no canopy. They were sailing from Maine to Florida and they found the sea warmer than the wind. Consequently they spent much of their time in the water, hanging onto the dinghy. The problem was, they were in shark infested waters! Only two survived.

During World War II, the rafts used did not have canopies and many people were found dead due to the wind and cold.

The right sort of canopy is very important. There are two basic types; the self-erecting and the type which requires manual erection.

As with everything in survival, every good point is tempered with a bad one and so a sacrifice has to be made somewhere. There are good and bad points in both these types of canopies but the good in the self-erecting canopy by far outweighs the bad.

Imagine you have been forced to evacuate your aircraft into cold, rough seas. The wind is blowing hard. Having managed to get to the raft, you then have to find the separate canopy, get it out despite the wind and erect it with poles. In reality, in these conditions, this will not work. The canopy will be lost or torn and the occupants in the raft will be too shocked and cold to manage this task.

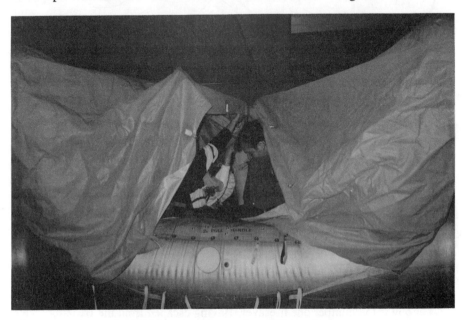

**Cabin crew fasten the self-erecting
canopy of their raft in a training session.**

The benefits of an automatic canopy cannot be stressed too much. The canopy's shape is formed by inflatable tubes and so is at least partially erected from the moment the raft inflates. It is also attached to the raft so it cannot be blown away or lost. The openings can then be closed very easily just by closing the Velcro fastenings or lacing the doors together.

The one negative point about a canopy such as this is that it can pose problems for helicopter rescue. It is possible for the downdraft of an overhead helicopter to flip the raft over and even blow it several hundred yards along the water's surface. However, the important factor is being able to open entrances enough, on both sides, so that this downdraft blows straight through the raft and doesn't make it unstable. The solution, then, is to make sure that self-erecting canopies can be opened on both sides, or that the canopy can be collapsed altogether, by letting air out of the canopy tubes.

As long as this basic rule is applied, the benefits of the self-erecting canopy are unrivalled.

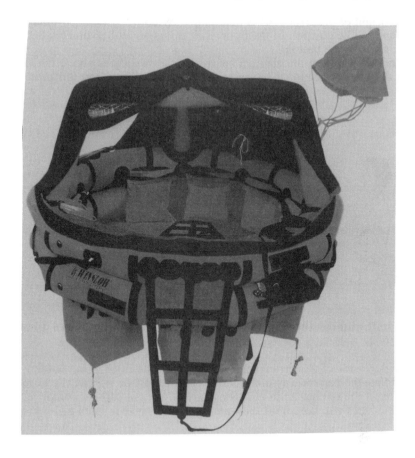

Raft with self-erecting inflatable canopy.
This raft has an inflatable canopy which erects automatically as the raft
inflates. The canopy can be opened up on both sides of the raft for safety
when a helicopter is in the vicinity.

Inflatable Floor

A good inflatable floor will provide a huge amount of insulation between the
water and the occupants of the raft. This can make it considerably warmer,
helping to minimise the risk of hypothermia. In hot climates, this floor can be
deflated to help cool the raft occupants.

Freeboard

This term refers to the height from the floor to the top of the tubes. The higher the
sides of the raft, the greater the protection offered to the occupants and the more
comfortable the ride.

The height of the freeboard is determined by both the size and number of the tubes and by the position of the floor. Raft floors are either situated between the two tubes or at the bottom.

It can be seen from the diagram below that the two tubes above the floor give by far the most protection. The larger the tubes, too, the greater this protection.

Raft with one tube.

Raft with two tubes.

The downside of this arrangement is that with the floor at the bottom, the raft must inflate the correct way up. If it does not it must be righted (See Chapter 7). However, if the raft has an inflatable canopy, it can not capsize completely as the inflatable tubes will keep it off the water. This will make it much easier to right.

**Capsized raft with inflatable canopy. It can be seen that
a raft designed like this will never capsize completely.**

Drogue and Ballast Pockets

These are both important features as their efficiency will determine the stability of the raft. The drogue is the sea anchor which slows drift and which helps prevent capsize. It should be as big as possible.

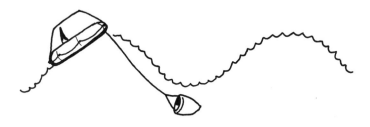

The drogue on a liferaft.
A good drogue should serve to pull the raft back from the
point at the crest of the wave where it is about to capsize.

It is the size of the ballast bags, however, that really determines stability. The largest ballast bags will fill with the most water, making it almost impossible for the raft to be capsized.

Those rafts that are made with the floor in the middle of the tube; that can be used either way up; will not have ballast bags and this is a serious flaw. Many of these rafts rely on a suction system to keep them stable but in reality this is not nearly efficient enough. It is not only the waves that can capsize a raft; the smallest corner that is exposed above the water will be picked up by the wind resulting in the raft being blown over.

Boarding System

Boarding the raft from the water is one of the most difficult things to do and so the better the boarding system the easier this will be. This is very important as the aim is to get out of the water as quickly as possible.

The raft's painter must go directly to the boarding ladder as it is vital that the swimmer does not have to let go of the line once holding it, especially in cold conditions.

Entry systems are usually designed as ladders or nets. Either works well but it helps to have a line continuing right into the raft that can enable the person entering to pull themselves right in.

Lifeline

The lifeline should extend right round the raft, allowing swimmers in the water to hold on to the raft at any point.

Equipment Bags

These should be large and plentiful inside the raft. They should also be able to be secured shut as keeping any useful equipment safe is paramount to both survival and morale.

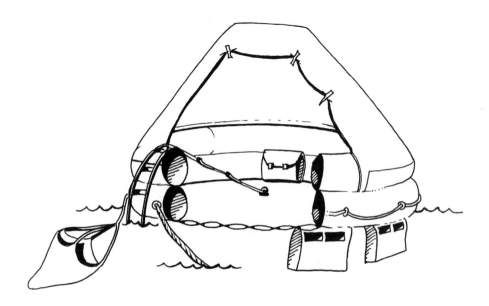

A raft with all the ideal features.

SURVIVAL EQUIPMENT

Emergency Locator Transmitter

These will be covered in more depth in the Search and Rescue chapter. However, a good Emergency Locator Transmitter (ELT) will be one of the most valuable things in getting yourself found. Although ELTs are fitted in most commercial passenger aircraft and in some light aircraft, it is important to have one in the raft as well, as those fitted in the aircraft are often destroyed on impact and the raft will not stay close to the aircraft for very long so even if it does survive and is working, it will lead rescuers to the accident site, not to the raft and survivors which will have drifted away from the crash site.

Water

This should not be touched for 24 hours except by those who are in urgent need of having fluids replaced; for example those who have suffered severe vomiting, injury or burns. When the water rations are opened, it is vital that they are shared out equally. Care should be taken with the storage of these in the raft as bags of water can split open very easily.

Water Maker

A reverse osmosis pump is the easiest method of creating water. A pipe is dropped into the water which sucks up sea water into the pump. This water is then pumped through a membrane which prevents the salt from passing through. The water collected in the bag, via another pipe, is then safe to drink. This is a sure proof method of collecting safe water but takes a lot of work for a relatively small amount of water. If raft occupants take it in turns to pump, however, you are assured of a constant supply of water.

The other method uses chemical de-salination kits but these are becoming rare as the reverse osmosis pump is more efficient.

Knife

A knife is essential equipment for any survival attempt but be sure to guard it from deranged survivors! Also be aware that it does not puncture the raft accidentally as this is a very easy thing to happen. It is useful if the knife has more than one blade and if these blades lock. There have been many serious injuries caused by knife blades closing on people while they are trying to use them. Remember that you will not be working in the calmest of conditions. Many survivors have also praised the benefits of being able to open the knife with only one hand; the other hand being busy holding onto something in rough conditions. The knife has been rated by those who have been in survival situations as one of the most valuable pieces of equipment.

Pump

This is important if you have an inflatable floor in your raft as you will have to pump this up yourself for added insulation. As well as the floor, the pump can be used to increase the air in the raft tubes if it begins to seep after a long time at sea.

Bailer Bucket and Sponge

Usually made of canvas, the bailer bucket is essential for emptying the raft of cold water. The sponge can also be used to help in this task. It is important to get the inside of the raft as dry as possible to help the occupants warm up.

Repair Kit

This should come with the raft and is important as it is the only method you will have of repairing any punctures or tears that occur. Repair items usually include patches and glue or plugs. The irony of the patches is that they require the tubes to be dry before the glue is applied! Plugs are often more useful as they hold better but are designed for larger tears. Remember, there is nothing wrong with making a hole bigger in order to fit a plug into it rather than managing with unsatisfactory patches.

A UK based company, Icarus, is now manufacturing specially strengthened raft tubes with tears for the very purpose of training crews in repair techniques.

Paddles

These can be used to move the raft away from danger or towards swimmers in the water. They can also be used to turn the raft and could act as a directional rudder. However, they are not going to be much use against strong currents and winds. As with every other piece of equipment, however, many uses could be found for them as necessary, including giving unwelcome sea creatures such as sharks a good smack on the back.

Flashlight

The obvious use for this is for seeing in the dark but maybe a more useful task of the flashlight is as a signalling device. A good light will be seen 20 miles away from the air. Make sure spare batteries are carried and do not waste these unnecessarily.

First Aid Kit

This will be covered in more depth in the ditching first aid section. However, two important items that should be included are sea sickness pills and sun tan lotion.

Survival Blankets

If immersion suits are not carried these can be lifesavers, even if used just for injured, very young or elderly people in the raft. They will help preserve body heat once out of the water.

Flares

These will be covered in detail in the Signalling and Rescue chapter. There are generally two types; hand-held and parachute flares. These should be stored carefully so they are kept dry.

Signalling Mirror

This is arguably one of the most useful items for attracting attention. It can be seen over 100 miles away by an aircraft at 25,000 feet and can be used as long as the sun is shining.

Sea Dye Marker

Another signalling device. This spreads bright green marker around the raft which can be seen from the air. This is an important piece of equipment as it is extremely difficult for the crew of a Search & Rescue aircraft to spot a raft in rough seas.

Chemical Sticks

These are very useful signalling aids. They are simply sticks containing light reactive chemicals. When you crack the sticks, there is a chemical mix which causes them to glow very brightly. They last for a considerable length of time and can be seen easily, especially at night.

Dental Floss

Dental floss can be useful for many things as it is very strong. Use it for tying things together or mending equipment. It is the strongest, most lightweight substance that can be carried for this purpose.

Decontamination Tablets

It is always useful to have these in a survival kit to purify water, even if the water has been pumped direct from the sea. Any illness caused by drinking disagreeable water will cause vomiting or diarrhoea, leading to severe fluid loss which must be avoided at all costs.

Fishing Kit

These are supplied in most survival kits and are very useful for many things. Unless the survival experience is prolonged, ie in excess of three days, there is no need to worry about food but the components of a fishing kit can have many uses.

Pliers

The more tools that are available to you the better as you never know what problems you are going to have to solve.

Each raft will come with a survival kit. Make sure you know what is in it and if you do not think its contents are sufficient, ask the raft manufacturer to add the items you want. Alternatively, you can do this yourself. It is imperative that your kit is complete and that it is always stored in, or attached to, the raft. It is no good putting these extras in the aircraft because the chances are, that is where they will stay.

Choose your raft carefully and do not skimp on design because of money. Obviously money will be a factor but when you are buying, imagine the conditions you will be dealing with if the worst happens. Your raft may very well be the only thing that saves your life!

CHECKING EQUIPMENT

It is imperative that survival equipment, especially rafts and lifejackets, are inspected and serviced in accordance with manufacturer's regulations. There have been many incidents in which it has been assumed that survival equipment will work when in fact that equipment has been packed away for years without having ever been checked. Needless to say, much of this equipment has failed when lives have depended on it.

9   First Aid at Sea

The injuries that are going to occur through a ditching will be numerous and potentially very serious. There will be the injuries sustained from the impact - these could be anything from cuts and bruises to broken bones, burns, crush and whiplash injuries, and there will be the injuries and complaints sustained later on from the effects of the water and the cold. These could include drowning and near-drowning, hypothermia, frost bite, heatstroke and sun burn in hot climates, sores from sustained time in the raft and psychological problems from the nature of the situation. Sickness and shock will also be prevalent.

It must be understood that, although as the crew you are responsible for treating these injuries, you are also responsible for yourself and the welfare of everyone else. By risking yourself to treat a few serious injuries, you could impede your own survival and thereby that of other survivors. Remember, the passengers are looking to the crew for their own salvation. It is important that you survive.

Do what you can but while in the water, the priority must be to get to the rafts, if possible and get out of the water. This is going to be, to a degree, an 'each man to himself' task. Once you are in the raft yourself, you can then begin to help others more effectively because you are no longer fighting for your own life in the sea. You can try to resuscitate a person while in the water, for example, but it is better to get them into the raft. While you are both in the water you are both risking drowning and hypothermia. The victim may die anyway - or may already be dead - and you cannot risk going with him or her.

The best policy is to help the sick and injured to the rafts and get them out of immediate danger as quickly as possible. Then worry about treating them or trying to revive them.

Drowning

This is one of the most immediate problems. Drowning can occur in minutes, even in cases where a life jacket is correctly worn, unless the life jacket has a spray hood fitted.

In the water, try to keep the casualty face up so their nose and mouth are free from the water. Keep their head tilted back to help them to breath. If resuscitation is going to be attempted, this should be in the form of mouth to nose in the water - never mouth to mouth. By opening their mouth to do this you are increasing the risk of water entering the airway and increasing the problem. Also, if you carry out mouth to mouth resuscitation, you must pinch the nose to shut off this airway. You do not have enough hands to do this and keep yourself afloat while in the water. You cannot carry out chest compressions in the water.

The unconscious person should be monitored as soon as they are in the raft, to see if they have a pulse and if they are breathing. This should be checked at the Carotid Artery at the neck. If neither breathing not pulse can be detected, Cardio Pulmonary Resuscitation (CPR) should be started. Ensure that chest compressions are not given until absolutely sure the heart has stopped.

Before giving mouth to mouth resuscitation, open the casualty's mouth and check it for debris that could impede the entry of air into the lungs or even be pushed further down the airway by air that you will be breathing in.

Tilt the head backwards to open the airway, pinch the nose tightly shut and open the mouth wide. Give two breaths before starting chest compressions, if appropriate. To find the correct place for compressions, locate the bottom of the sternum (or breast bone), put two fingers below this and then place the heel of your hand below this point, interlocking the other hand into it. All crew should be conversant with first aid and CPR and so should have practised this at length in each recurrent class. Obviously in a raft, space constraints are going to cause problems.

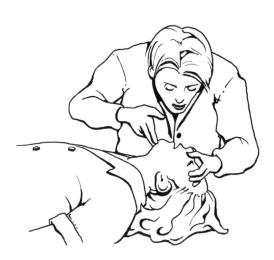

Administering Cardio Pulmonary Resuscitation to a casualty.

Try to put something solid under the casualty, if there is anything available, although this will be unlikely. Compressions made on a raft floor will not be very useful as the rubber floor will give under the pressure.

Compressions should be made at the rate of 100 per minute. If one person is carrying out CPR alone, then the sequence should be one breath to five compressions. If there are two people available for the task, the sequence should be two breaths to 15 compressions. The person carrying out the breaths is responsible for monitoring the casualty's condition. Check, every so often, to see if the casualty's heart has started again. Use mouth to mouth resuscitation in the raft unless there are poisons around the mouth or there is severe facial injury. In these cases, mouth to nose resuscitation is safer.

Technically, resuscitation should not be stopped until either the heart starts and the person begins to breath on their own or until medical help arrives. In a survival situation, however, where the CPR of a casualty is just one job among dozens needed to keep everyone alive and the raft managed, you have to be sensible. The chances of the person recovering become slimmer as each minute passes and you have to consider your condition, and that of the other occupants of the raft, too. You could be more use making sure that water is being pumped, the raft repaired, a look out watch kept and that the knives are removed from the grasp of someone who is beginning to behave unpredictably. It will come to a point where the safest thing will be to give up and to get rid of the body. This may sound harsh but here we are talking about the survival of the fittest.

If the person does show signs of recovery, turn them onto their side, in the recovery position as much as is possible in a raft and keep them warm. As the organs of the body naturally fall to the left hand side, it is safer to put the casualty on this side, unless serious injuries dictate otherwise. Monitor the casualty regularly.

The recovery position.

For those who are conscious from the beginning, do not be deceived - they may have inhaled water and this may not be immediately apparent. Coughing can help remove the water from the airway but be aware that these people may deteriorate suddenly at any time. Signs to watch out for are chest pain, difficulty in breathing and a bluish colour around the lips and face. Try to encourage people in this situation to breath deeply to try and restore oxygen levels.

Be aware of secondary drowning. This is described in more detail in Chapter 6. All those who have been in the water should attend a hospital for a chest x-ray and examination by a doctor once rescued.

Understanding Hypothermia

Along with drowning, this is the other quickest killer. In the water it can kill in two ways; directly and by incapacitating a person to such a degree that they will be overcome and will drown.

There are four stages of hypothermia as associated with water immersion:

Initial Immersion - the first couple of minutes causing 'cold shock' reactions.

Short Term Immersion - between three and 15 minutes causing exhaustion and drowning.

Long Term Immersion - in excess of 30 minutes causing progressive cooling.

Post Immersion - causing collapse and secondary drowning. This normally applies to about 20% of all casualties.

Initial Immersion produces a dramatic increase in heart rate, breathing rate and blood pressure. These responses will decline after a few minutes of immersion but have been known to incapacitate victims, even to the point of death. They should not be taken lightly. Middle aged or elderly people will suffer most, especially if they already suffer from high blood pressure or heart disease. Actions to prevent serious problems include entering the water slowly rather than just jumping in, giving the body more time to adjust to the severe drop in temperature and then keeping still for a few minutes once in the water to stabilise the body. This will obviously depend on the proximity of danger and rescue.

Short Term Immersion will exhaust even usually strong swimmers very quickly. This is because it is difficult to maintain adequate arm and leg strokes when fighting for breath or suffering from rapid gasping - part of the 'cold shock reaction'. It is easy for such casualties to inhale water and drown. This stresses the importance of lifejackets that will keep unconscious bodies facing upwards - those wearing lifejackets with spray hoods stand even more chance.

Long Term Immersion can cause hypothermia very quickly. The principle symptom of this is loss of body heat. The normal body temperature is 36.9 degrees. If the temperature falls to below 35 degrees, that person is said to be hypothermic. The first signs are shivering and increasing numbness of the body, especially the extremities, as the blood goes inwards to warm the vital organs to keep the body alive. Increasingly sluggish reactions will then occur, accompanied by slurring of speech and vision difficulties. Irrational and irritable behaviour may also be present, as may cramp, nausea and, later on, complete lethargy.

Gradually, shivering will be replaced by persistent muscular rigidity and soon after this the casualty will become unconscious with depressed breathing and a slow pulse rate. They will be cold to the touch, very pale and may have dilated pupils. As the body temperature drops to about 24 degrees, the casualty will die of heart failure.

The timescale for these events depends on many things; the water temperature the person has been exposed to, the wind chill effect and the drop in temperature that this causes, the amount and type of clothing worn, the age and physical fitness of the person concerned, the amount of body fat present, the amount of movement undertaken while in the water and the corresponding amount of heat loss from the body.

Treating Hypothermia in a Ditching Situation

Remove wet clothing as quickly as possible and replace it with dry clothing, if any is available. Do this unless the casualty is wearing natural fibres that will warm when insulated, and can be wrapped in a survival blanket, or similar, to guarantee warmth. Warm the person by any means - body heat is one of the best insulations so huddle close and hold them. Remember to cover the head as it is one of the greatest areas of heat loss.

Warm the casualty from the core outwards. Do not use direct sources of heat on the patient or warm up the extremities rather than the core of the body as anything that is likely to draw the blood back into the limbs will remove it from the vital organs where it is keeping the person alive. Keep the casualty still; do not administer alcohol and do not rub the limbs.

Another reason why the cold limbs should not be warmed up too quickly is that they will have been starved of oxygen and so may have developed poisonous toxins that will then be released back into the core of the body.

Hypothermic casualties can appear lifeless for a long time before showing signs of recovery so if the person is unconscious, keep the airway clear and try mouth to mouth resuscitation. Be aware that they may be alive even though they look very dead and they may take some considerable time to show any sign of recovery. Do not give chest compressions until you have been checking the heart constantly for well over a minute. The heart of a hypothermic casualty may well be beating although it will have slowed down so much that you will barely be able to detect it. Giving chest compressions if there is an output will kill.

It can be very difficult to know whether a person is suffering from drowning or hypothermia or both. Remember that the recovery rate from drowning is much more successful if the person has been in cold water because the body systems shut down that much more when very cold. Even if you suspect drowning as the reason for their condition, then, persevere with treatment as long as you can safely do so - bearing in mind the many other survival factors you will be expected to cope with.

Once the hypothermic person is in the raft, encourage the blood to flow to the heart, lungs and brain by lying the casualty down in a slightly head down posture - in the recovery position but with the legs slightly raised.

Be aware when pulling the hypothermic casualty out of the water that they should not be lifted out vertically but horizontally. This may be considered academic in a survival situation where you have to use any method you can to get someone into the raft, but be aware of it and if you can use two people to lift them in horizontally, it may well save their life. When being rescued by helicopter from the water, the rescuers will always take this precaution.

The reason is that lifting a casualty vertically from cold water can cause a 30% reduction to the heart output which may result in collapse or even quick death. While in the sea, the blood pressure in the body is lessened but this is not felt as acutely while in the water because of the compensating pressure of the surrounding water on the body. Quick, vertical removal of the patient from this environment can cause the problems explained above because the blood will flow down to the lower limbs, leaving the vital organs without an adequate oxygen supply and without any body heat. If the casualty is lifted horizontally, the blood is allowed to remain around the vital organs such as the heart, lungs and brain. The condition is known as hyperbaric shock.

Keeping the casualty warm is the key to success with this complaint. Wrap them in any spare clothing or survival blankets and monitor them. Put them between other raft occupants for warmth.

Be aware that people may suddenly go hypothermic after some time in the raft. Do not assume that because they were not a serious case at the beginning that they will not suddenly become so. Encourage everyone to huddle together, to put on any spare clothing, to sit on anything that will keep them off the raft floor, including their lifejacket if it is very calm, and to keep moving their fingers and toes to maintain circulation.

Watch for the quiet person who will probably not complain about their plight. They will gradually slow down and become woozy as if drunk. If no-one is on the look out they could be unconscious before anyone has noticed!

After-effects are very common for anyone who has been very cold. These can include collapse, secondary drowning, pneumonia, chest infections, swelling in the brain and other problems due to lack of oxygen. Again, all survivors must be given a thorough medical examination once safe.

Cramp

This is another common feature of cold water immersion. It will immobilise even the strongest swimmers very quickly and is brought on because of an increased respiratory rate which causes hyperventilation. The greater the number of shallow breaths taken into the body, the greater the exchange of gases and amount of Carbon Dioxide left in the bloodstream. It is this Carbon Dioxide that causes cramp in the muscles.

Burns

The treatment for burns is covered in the Land Survival chapter. In a ditching situation, take advantage of the water around you to cool the burn. If possible, simply submerge the burn in the sea. If not, wet cloths and keep the offending area cool.

Cover the burn with a clean dressing and put this person on the priority list for water. Burns dehydrate and the kidneys will be at risk if fluid is not replaced. Give water in small quantities by mouth.

Sun burn also comes under this category. Although this may sound trivial in comparison, if you are out in very hot climates with little protection, especially surrounded by water which reflects the sun's rays, the sun can cause very serious burns. Use sun cream, if available, if out in even remotely warm conditions as the sun is very misleading and can burn far quicker than you think, even through thick cloud. Keep skin covered.

Bleeding

This, too, should be treated as it would be in a normal first aid situation on land. Take advantage in a sea survival situation of having salt water that will be helpful in cleansing more minor wounds.

Be aware of internal bleeding - it is quiet, easy to go unnoticed and can be deadly! The principle symptom is shock which increases in severity as time passes. Treat for shock and monitor at all times; unconsciousness can occur very suddenly.

Always be aware of the dangers of blood borne pathogens.

Frostbite

Frostbite will be prevalent in a sea survival situation. See chapter 16 for signs and treatment.

Urine Retention

Be aware of this and encourage everyone to pass water within two hours of entering the raft. If this does not happen, a mental block may set in which will prevent people from being able to pass water. This may lead to problems with the kidneys because toxins are not being removed from the body. Urine production will be reduced after a time by water rationing.

Vomiting

This will be one of the worst problems you will have to deal with. It is very unpleasant and although not serious on the surface, vomiting leads to severe fluid loss which is critical if you do not have water to replace them.

There is nothing much you can do about it except hand out the sea sickness pills as soon as possible - preferably about 20 minutes before ditching - and insist that everyone takes them. Hand out more later on. Use sick bags to prevent the chain reaction of every one vomiting into the raft and use these as insulation.

Heat Exhaustion and Heat Stroke

Both these conditions are covered in Chapter 16. In a raft, you have the added benefit of being surrounded by water that you can use to cool the affected person down. Cooling can be achieved by wetting clothes, splashing water over the patient and by opening up the canopy to let wind through but do not collapse it completely because you need it to protect against the sun's rays. Wetting the outer sides of the canopy can also help cool the interior. Dehydration is also covered in Chapter 16.

10 Signalling and Rescue at Sea

Your priorities will vary throughout your survival experience according to the situation you are in but despite everything else, you have one ultimate aim and that is to be rescued.

The fight is always to maintain life in order that you, your crew and passengers can stay alive to be rescued. This is important but rescue itself must always be at the forefront of your mind, whatever other pressing priorities loom in. The quicker rescue comes, the easier the survival task is.

There are two parts to the rescue story. One is letting people know that you are in trouble in the first place and giving them some idea of where to look for you and the other is the actual rescue operation. This may sound easy but many a 'survivor' has given up too early and died during the rescue attempt. Being aware of how the rescue will take place so that you can help with it as much as you can, will make it happen smoothly and quickly.

Signalling

This process is crucial as, contrary to popular belief, you are unlikely to be traced to your impact point and rescued within half an hour by someone who has nothing better to do than spend their days tracking your particular aircraft.

Signalling should begin before impact. Radio calls can be made, giving your position as accurately as possible; aircraft transponders should be set onto the emergency frequency of 7700 and emergency beacons should be switched on.

This tells the authorities that an aircraft is in trouble and gives them an idea of where that aircraft is going to make an emergency landing. Having thus alerted the authorities to the fact that you are in the plight that you are, all efforts must then go towards being found. In the sea, either as a swimmer or in a raft, you will not stay in one place for long. The drift can be astounding and if it takes a rescue aircraft one hour to get to where your aircraft went in (and this is an abnormally quick response - depending on where you are in the world it could take up to three days) you could be miles away from that spot by the time rescue teams get there.

The rescue services are accustomed to plotting drift and wind speed on a graph in order to trace survivors who are afloat so all is not lost but remember that this takes time and while they are following your progress, you are drifting further and further away.

There is then, of course, the problem of actually finding you - a liferaft bobbing in the sea looks like a pin prick from 500 feet, if you can see it at all in a rough sea - and this is when constant signalling and making yourself visible comes in.

Keeping a look out from the raft is one of the basic rules of raft management. Not only are you looking for danger and other possible survivors in the sea, you are also looking for signs of rescue. This is essential because even though an aircraft can accurately 'home in' on an emergency beacon that is emitting on the International Distress frequency of 121.5 MHz, you can guarantee it does not have visual contact with you and that is when using signalling equipment such as mirrors, flares, sea dye marker and smoke comes in to enable it to actually find you.

If you miss these rescuers who are out searching for you, you will miss the chance to let them know that you are there. If they are searching a large area and cannot see you in one place, they will soon move on to search another area. You have then lost your chance. By the time they come back to where you are, it may be too late!

Two recent incidents can be used to show how difficult it is to be seen in the sea. One aircraft, ditched in the Atlantic Ocean, left five people clinging to a liferaft. Rescue aircraft from the US Coast Guard were out searching almost immediately. Six aircraft were deployed and searched the area in which the survivors were waiting for over five hours before they actually spotted the raft.

A member of the Coast Guard told how frustrating the situation was. "We knew they were there but we just could not see them." The survivors later said that the aircraft had passed right by them on numerous occasions and they could not understand why they had not been spotted.

Another ditching in the Atlantic highlights the same problem. A light aircraft ditched leaving a father and son afloat in the sea. The rescue aircraft flew right overhead many times but the crew were totally unable to see the survivors.

Emergency Locator Beacons

One of the most useful tools of signalling and survival, and one no aircraft should take-off without, is the Emergency Locator Beacon. There are three types of beacon:

PLB - Personal Locator Beacon. This is a small, hand held beacon. This sort of device is designed to be carried out of the aircraft and into the raft to ensure that a beacon stays with the survivors at all times. A beacon in an aircraft that is getting further and further away from you as you drift is not very useful and so it is essential, in a ditching situation, that PLBs are carried and that they are taken from the aircraft, if at all possible, in an emergency. This item should be on the priority list of things to take.

Most raft manufacturers also put a PLB in the raft's emergency equipment bag and this can be a life saver on those occasions when it is not possible to retrieve anything from the aircraft. All rafts should be equipped with such a device. Always remember that you can not have enough emergency beacons. The other time when PLBs become essential equipment is for those occasions, unfortunately all too frequent, when the aircraft's own beacon has been damaged by the impact force.

PLBs are activated by manually switching them on. Some of these devices also have a two-way radio function built into them to enable survivors to talk to rescue authorities.

ELT - Emergency Locator Transmitter. This is the term given to aircraft fixed beacons. Usually fitted in the tail of the aircraft, these beacons can be activated manually, either by a crewmember - usually from the cockpit but sometimes from the cabin, or by a specified g force on impact.

There have been incidents in which an aircraft's ELT has been damaged during impact. It seems that one of the weaknesses of these devices stems from the fact that they are designed to be remotely controlled. If there is a capability, for example, for the Captain to switch the device on from the cockpit, then wires must run from the device all the way to the cockpit at the other end of the aircraft. This increases the risk of damage to these wires which then isolates the device.

The ideal situation is for aircraft to have an ELT fitted but also for the crew to carry PLBs. In any situation in which the survivors move away from the aircraft, almost bound to happen in a water survival situation, these PLBs will become a life line.

EPIRB - Emergency Positioning Indicator Radio Beacon. This is the name given to the emergency beacons carried on boats. These are usually activated either manually or by immersion in water.

How do PLBs/ELTs Work?

All PLBs and ELTs work by transmitting a radio frequency which can be used to provide the location of the beacon. A system known as COSPAS-SARSAT, which operates using a satellite monitoring system, is used to pick up these signals and relay them back to earth.

All civilian beacons transmit on a frequency of either 121.5 Megahertz (MHz) or 406.025 MHz. Many of those transmitting on a 406.025 MHz frequency also transmit on 121.5 MHz.

The older 121.5 MHz beacons are more common and cheaper to buy but are less efficient than the more modern 406.025 MHz beacons because of their reduced search range.

121.5 MHz Beacons

These are constrained by providing a location accuracy of up to 20 square miles although once the beacon has been located, it does allow the rescuer to 'home in' on it for an accurate rescue. There is no identification transmission capability with the 121.5 MHz frequency; the beacon is not capable of data encoding and so there is no way of knowing who the beacon belongs to, or what type of craft is in distress, as there is with the 406 MHz beacons.

The other problem with this type of beacon is that the majority of emergency transmissions are false alarms.

406.025 MHz beacons

The location accuracy of these beacons, usually referred to as 406 MHz beacons, can be as small as one square mile and the signal contains digital information unique to each beacon which provides a link to information contained in a registration database. This registration information can identify the aircraft in distress, thus quickening response time. There are far fewer false alarms with these beacons.

The 406 MHz frequency is not as suitable for use as a homing signal and it is for this reason that many 406 MHz beacons also transmit on a 121.5 MHz frequency. Other manufacturers have incorporated a homing device into their 406 beacons.

COSPAS-SARSAT System

Every polar orbiting and geostationary satellite carries a Search and Rescue Satellite-Aided Tracking (SARSAT) payload that can detect and locate emergency beacons. These SARSAT payloads are provided by Canada and France while Russia operates very similar equipment on navigation satellites known as COSPAS.

Both are used in an international, co-operative Search and Rescue effort entitled COSPAS-SARSAT.

Polar Orbiting Satellites

SARSAT polar orbiting satellites circle the Earth every 102 minutes at an altitude of 528 miles. COSPAS satellites, however, circle the Earth every 105 minutes at an altitude of 620 miles. Both of these orbiting satellites travel at a speed of 15,500 miles an hour and can view an area of over 2,500 miles in diameter as they orbit the Earth.

The satellites overfly the poles on every orbit and so coverage of emergency beacons is best in these areas and least at the Equator. In between these two areas, in the mid-latitudes, the average waiting time for a satellite pass is 30 - 45 minutes.

Geostationary Satellites

These satellites orbit the Earth at an altitude of about 22,320 miles above the Equator. They carry 406 MHz receivers and because of their high orbit, can see large areas of the earth's surface continuously.

The COSPAS-SARSAT system consists of a network of satellites, ground stations, mission control centres and rescue co-ordination centres.

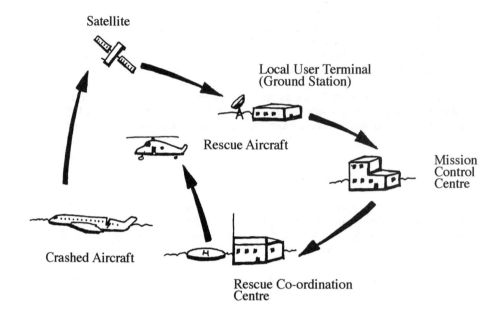

Satellite

Local User Terminal
(Ground Station)

Rescue Aircraft

Mission
Control
Centre

Crashed Aircraft

Rescue Co-ordination
Centre

The five sections of the COSPAS-SARSAT system.

When an emergency beacon is activated, the signal is received by a satellite and relayed to the nearest available ground station. This ground station, known as a Local User Terminal, will process the signal and calculate the position from which it originated. This position is then transmitted to a Mission Control Centre where any identification data from the system database is added to the information on that beacon.

From here, the Mission Control Centre will transmit an alert message to the best placed rescue co-ordination centre to respond to an emergency in that location. If the location of the beacon is in another country's service area, the alert is transmitted to the Mission Control Centre in that country.

Emergency Beacon False Alarms

The vast majority of emergency transmissions are false alarms, thus wasting huge amounts of valuable search and rescue time that could be used for a real emergency. With the 121.5 MHz frequency, the COSPAS-SARSAT system receives on average 1,000 false alarms for every real emergency! This is greatly reduced with the use of a 406 MHz beacon; only about eight false alarms for each real emergency. In addition, because of the identification capability of the 406 MHz beacons, about 80% of 406 MHz false alarms are resolved very simply by a telephone call.

The following advice should be taken by those using a 121.5 MHz beacon to reduce false alarms to a minimum:

* Ensure the beacon is properly mounted.

* Maintain fresh batteries in accordance with the manufacturer's recommendations.

* Ensure the battery is disconnected when the unit is being shipped or disposed of.

* Ensure you know how to use the beacon - before an emergency occurs.

* Monitor 121.5 MHz after each landing to ensure that your ELT is not accidentally transmitting.

* When testing the 121.5 MHz beacon, do so only during the first five minutes of any hour and limit the transmission to three audio sweeps.

The following advice is for users of 406 MHz beacons:

* Register the beacon.

* Mount the registration decal so it can easily be seen without removing the ELT from its bracket.

* Test the device as recommended by the manufacturer.

Beacons are available that will operate between the temperature ranges of -40 degrees C to +55 degrees C for a period of not less than 48 hours.

It is a false economy to buy cheaply when lives are at stake in a survival situation. Unfortunately, this point is never driven home until the unthinkable happens.

Signalling

The ELT or PLB is one of the most useful items for letting people know that you are in trouble and for giving them an idea of your location. Turn the beacon on and leave it on for the first few hours of any survival situation. However, if the ordeal becomes prolonged and it is obvious that you are not going to be found quickly, preserve the life of the beacon's batteries and only switch it on intermittently or when you think there is the best chance of someone picking up the signal, ie when aircraft fly overhead or if you sight ships. Remember that once the batteries are dead, you no longer have an emergency beacon. If you are sure that someone has been alerted to your plight and you think they are out looking for you, switch it on to help them 'home in' on you.

The key to signalling is not to waste resources when there is no-one to see them. Save your equipment and let it do the best job for you.

The majority of the signalling equipment you will carry will be more useful in helping people to find you once they know that you are in trouble and have a rough location in which to start looking.

Global Positioning Systems

These, especially when incorporating an emergency locator beacon, will allow rescuers to pinpoint your exact position and monitor it as it changes.

Cellular Telephones

These can be used, providing there is enough signal wherever you have come down, to summon help. It is possible to triangulate your position by using different masts.

Signalling Mirror

This is a very useful device as it can be seen for over 50 miles at 25,000 ft and can be used at any time when the sun is shining.

Using a signalling mirror.

To use it, put your arm up towards the object you want to signal, such as an aircraft, and keep this object between V-shaped fingers. Then, keeping the mirror at your shoulders, shine it so the central spot shows on your fingers. By doing this, you know that the glare of the mirror is aimed directly towards where you want it to be seen.

A signalling mirror should be an essential item in your survival kit. If you do not have one, however, all is not lost, you can improvise! Anything that shines will work as a signalling device, even if it is not quite as efficient as the real thing. The shiny part of a credit card has been used by a survivor in the water for example, and was seen easily enough by a passing aircraft that was not even aware that there was anyone in trouble. The survivor was picked up safe and well.

A signalling mirror is a great device and can be used to alert aircraft or ships of your predicament even when they are not out looking for you. If, as an aircraft pilot on a routine flight, for example, you saw such a signal flash from the water, you would be sufficiently alerted to report the incident and the geographical position to the authorities.

Sea Dye Marker

This is a dye which can be used to colour the sea around you a bright green colour. The dye can be seen from a great distance but does disperse after about three hours (quicker in rough seas) and so can not be relied on as a long term signalling device. The other draw back to it is that as well as dispersing, it also tends to linger where first dropped and so a raft that is drifting will leave it behind relatively quickly. It is corrosive and so should not be allowed to come into contact with the fabric of the raft. Tying it to the raft, therefore, should be avoided.

Flares

These come in two types: parachute flares and hand held flares. All flares should be used as location aids. Do not fire them until you are sure there is someone there to see you. Note that to work, all pyrotechnics must be kept dry.

If a rescue boat, for example, is firing off green, or at night, white flares, it means that they are looking for you - not that they have found you. This is your cue to use your flares, especially the hand held sort, to give them a visual position of where you are.

Parachute Flares

These are otherwise known as rocket flares. You fire them into the air and they shine bright red, rather like fireworks. They have a visibility approaching 28 miles. Make sure that you fire these flares well clear of the raft canopy and not in the direction of any other rafts. Fire them down wind and look away as you let them off so that any hot particles that fly away do not go in your eyes. DO NOT FIRE THEM IN CLOSE PROXIMITY OF AIRCRAFT.

Each make of flare will have slightly different operating instructions - make sure that you and your crew are totally familiar with how these flares are activated before you have to use them for real.

Hand held flares generally have two ends; they let off bright orange smoke from one end for use during the day and a red flare for use at night. There is a cap protecting both ends and one of these should be embossed so you can tell which end is which in the dark. Make sure everyone knows which is which.

To use these hand held flares, remove the cap and pull the short string that is attached. A sharp pull is normally required, and, as with any pyrotechnic, take care as you set it off because any hot particles that come loose can cause damage to both people and the raft.

Hand held flares get very hot as they burn. Lean right over the edge of the raft and hold the flare pointing directly out in front of you to prevent your hand being burnt if hot particles run down the side of the flare. Do not hold them it upright for this reason. If you do have to drop it, throw it into the water. DO NOT LET IT FALL INTO THE RAFT!

These hand held flares are excellent for enabling rescuers to see you once they are close but not yet in visual contact. In addition, the smoke end of the flare can be used to show rescue helicopter pilots the wind direction as they are positioning for a winch. Hand held flares will burn for an average of 18 seconds and can be seen from a distance of about three nautical miles.

The other type of flare often used is called a mini-flare and is like a light-weight pistol. The flare capsule is loaded into the pistol and it is fired into the air. It forms a bright orange ball. Take the same precautions with these flares as with parachute flares and again, do not use near aircraft.

Whistle

Aviation life jackets should all be fitted with a whistle. This is useful in attracting attention. The International Distress Signal is six short blasts, followed by one minute's silence before the next set of six short blasts. Although the whistle will not have a very long audible range, especially in high winds, it is a useful aid to attract attention to rescuers who may be close by, especially in the dark or in bad visibility.

Chemical Light Sticks

These are useful at night for helping rescuers see you. They are hand held and need only to be broken to let the chemicals mix, causing them to shine. They are very bright and last for a long time. Once the chemicals have mixed, however, the sticks cannot be put out and you must wait until they extinguish naturally.

SEE/RESCUE Streamer

This is a relatively new product and is simply a long, high-strength, bright orange polyethylene streamer. It is designed for use by any survivor, whether in the sea or on land and makes even a single person easily visible from the air. The device comes in different lengths; 25 foot and 40 foot and folds up into a small, lightweight package. It is probably one of the most useful pieces of equipment to allow rescuers to see you easily. Every raft and survival kit should contain a SEE/RESCUE streamer.

SEE/RESCUE streamer in the water.

Flashlight

This is another useful device which can be used to signal in the dark. It is especially useful for pin-pointing your exact position once a rescue boat or aircraft is in the vicinity.

Strobe

Most life rafts are equipped with strobe lights. They are normally fitted to the apex of the raft. Strobe lights flash a very bright, intermittent light and have an eight or nine hour life. It is also possible to get hand held strobes.

What You Can Do Yourself

There are certain things survivors can do to help themselves be seen. Improvise; use whatever is in your survival kit, for example, an emergency blanket or aluminium blanket, for visibility. These have been known to show up on radar. Tie rafts together, with enough room between them so that each can ride the crest of waves independently. This will ensure that even when one raft is hidden in the trough, one will be visible on the crest.

The greater the number of rafts tied together, the larger the mass there will be to be seen. Remember also, the rule that there are no straight lines or circles in nature - if you have a number of rafts tied together in a straight line, these will show up easier than individual rafts bobbing up and down.

Splashing also shows up, especially in calmer seas, either from the raft or from a swimmer in the water. In addition, people in the water should try to lie with as much of their body showing to the surface of the water as possible. A full body length is much easier to spot than a head. This position should only be assumed when a rescue aircraft is in the area. At all other times, conserve energy by keeping the body as small as possible.

RESCUE

Rescue, the ultimate aim throughout your ordeal, will come in one of three ways: by air, by boat or by drifting ashore.

If the aircraft goes down relatively close to shore, the most likely means of rescue will be by helicopter. If you are miles away from anywhere, however, aircraft can get out to you, can monitor your position and can even throw you supplies but you will have to wait for a ship to reach you to actually be rescued. This could take several days, depending on where you are and once the ship reaches you, rescue will occur either by helicopter, launched from the ship, or by small craft.

Take, for example, the three yachtsmen who got into trouble in waters half way between the south coast of Australia and the coast of Antarctica during a yacht race. Aircraft were sent out to the area in which the yachtsmen were in trouble while a ship was dispatched from Australia to carry out the actual rescue. It was a three and a half hour flight for the aircraft to get out to the area; it then had a fuel capacity sufficient to allow it to spend three and a half hours circling the distress area before it had to return to land to refuel. The aircraft was, however, able to drop Air Sea Rescue kits and radios to the yachtsmen.

It was first thing on Monday morning that the rescue services were alerted to the plight of the sailors. All three boats had two emergency beacons on-board each, all operating normally and so rescuers had a clear idea of the survivors' locations. Aircraft were sent out constantly from then on, dropping supplies and monitoring the position and condition of the men but it took the ship that had been dispatched to carry out the rescue, HMAS Adelaide, until Thursday to get to the scene! Imagine that your aircraft ran into trouble in a location a similar distance from land; do not imagine you will be picked up within hours.

Rescue may also occur by boat if you happen to be found by a passing one, or if rescue co-ordinators divert a near-by ship to go to your assistance. Again, if close to shore, there maybe a combined rescue attempt. In the UK, for example, the Coastguard and military helicopters, as well as the Royal National Lifeboat Institution (RNLI) boats often work together to perform search and rescue missions.

Drifting ashore is probably a rare form of rescue and even then, may not be rescue but merely a transition from one form of survival to another if the land which you drift onto is remote and sparsely populated. Strange as the idea of drifting ashore may seem, it must be remembered that Steven Callahan, who spent

76 days adrift and alone in a liferaft after his yacht capsized, came to his salvation by landing on the coast of a Caribbean island after drifting 1800 nautical miles.

Helping the Rescue Attempt

Rescue may seem like the final part of your good fortune but it is vital that you do not give up 'surviving' until you are well and truly safe. A bungled rescue attempt due to throwing away caution and common sense half an hour too early has left many a person dead. The more people can understand about how the rescue will be conducted, the more they can help their rescuers and the quicker and more efficient the whole business will be.

Rescue by Air

If rescue is conducted by a helicopter, you will be winched into the aircraft to safety. The helicopter pilot will spend some time checking out the area and preparing for the winch. He will hover into wind to do this and this is when use of the hand held smoke flare can help him tremendously.

It is important that you are ready to be winched and that you take some basic precautions in the raft. The first job is to collapse the canopy altogether if you can or if not, open it up, on both sides of the raft, as much as possible, to prevent the down draft from the helicopter capsizing the raft. Make sure the drogue is out to prevent drift and secure all loose items as they will be blown around violently.

Apart from this, you must wait and do exactly what you are told by the crew. The helicopter will first lower a winchman into the raft with you. Do not touch the winchman as he descends; helicopters build up an enormous amount of static electricity and the winchman will earth himself through whatever touches him first. Let this be the raft otherwise you, and he, will suffer a huge electric shock. Never attach the winch line to anything in the raft.

Follow the winchman's instructions. He will probably come into the raft and will send two survivors up at a time. Be ready to tell him of any serious injuries on-board the raft. It is important to re-distribute the load as the people move out of the raft. The lighter the raft, the easier it is for the downdraft to flip it over.

When being winched, you will have the strop under your arms and around your back. This is all that is holding you. Make sure the strop is tight, keep still and keep your arms down by your sides, or at least, if you must hold on, keep your elbows firmly down. The instinct to put your arms right up to hold on increases the danger of you falling through the strop. As you near the aircraft, just hang limply and let the crew manhandle you into the aircraft. Do not try to help them - they know what they are doing and will find it much easier without hindrance from someone who does not. Once in the aircraft, life jackets should be deflated as in all aircraft, just in case of a second ditching!

There have been reports of problems caused by inflated life jackets when winching. The strop pulls very hard on the back and problems can occur with breathing or with lifejackets being forced up and almost strangling the person being winched. This is an insoluble problem as winching has to take place and it is dangerous to deflate the lifejacket before reaching the safety of the aircraft in case there is a problem and the person being winched ends up in the sea. The best advice is to play it by ear and to make sure that life jackets are securely fitted, especially around the waist, so that they move as little as possible.

Remember too, that winching is getting you to safety and although it may be uncomfortable, or even feel dangerous, you are not in the strop for that long.

If you are being rescued from the water by a helicopter, much the same applies. The winchman will come down to you and will stay in the water until everyone has been winched to safety. Because of the risk of hyperbaric shock - the danger of lifting a long-submerged casualty from the water vertically (see Chapter 9), many Search and Rescue helicopter crews will send down a stretcher-like cage and lift each person horizontally.

SAR crews rescuing an injured casualty.
Search and Rescue helicopter crews will send down a stretcher-like cage to lift people who have been submerged in the water for any length of time.

Rescue from a Boat

This, in many ways, is more difficult as transfer from a raft or from the water into a boat can be tricky. If it is a lifeboat that comes to your assistance, do exactly what the crew tell you. Clear any debris away from the side of the raft, ie streamers or rope to enable the boat to come in close without getting entangled.

You may be given a large drogue to help keep the raft steady during rescue - use it, especially if this is a combined rescue and you are ultimately going to be winched to safety by air.

Be aware of stepping from the raft into a boat in high waves; if you don't judge it just right, the vertical distance between one and the other could be huge and you will end up in the middle. Make your move just when the crew say, otherwise you could jump just as the coxswain decides to turn away and try another approach.

Again, be prepared to tell the crew of any serious casualties who may need special assistance during the rescue.

In all cases, it is usual for rescuers to sink the rafts; it is too complicated to retrieve them and leaving them in the water may lead to false alarms.

Drifting Ashore

There will be signs that you are approaching land long before you actually see it. It is a common phenomenon when adrift to imagine seeing land every few hours when in fact, some cloud banks can look convincingly like mountain peaks.

Some signs include clusters of birds and bird sounds. While one or two indvidual birds are not a clear indication of land fall, clusters are. It is a usual pattern for birds to fly from land during the day and back towards land to roost. If you can hear the constant sound of bird cries, this is a sign that land is quite close.

Debris in the sea can also indicate land. Debris can be swept for hundreds of miles across the ocean but an increase in it, or in some form of vegetation, for example, coconuts, can be a sign that land is close.

Cumulus clouds that gather in one point in an otherwise clear sky can also indicate land as they tend to gather above high points, such as hills. A change in sea colour, too, can be an indication. Water that is coloured with silt, for example, has clearly come from a nearby estuary or beach.

When approaching land, try to attract attention by using, for example, emergency locator beacons, flares and signalling mirrors. If this does not work and the raft is approaching the shores, be aware of current, rocks, cliffs and reefs. To an extent, the occupants of a drifting raft are powerless to change direction but some leeway may-be gained by getting out the paddles and rowing to try and avoid such dangers.

Land Survival

11 Introduction - Land Survival

In the chapters on land survival, this book is really referring to extreme survival situations in the desert, jungle and Arctic regions, environments that cover huge amounts of the earth's surface. These are being covered as the 'worst case' scenarios and it should be realised that threatening environments can be found in areas much less extreme than these. The rule that this book is adopting is that knowledge about the very extremes will enable aircrew to cope in any survival situation that they might find themselves in.

The following chapters will consider protection against freezing temperatures, drifting snow, strong winds, scorching heat, incessant sun, humidity, lack of water and plagues of insects. It will cover skills such as personal protection, building shelters, procuring water, lighting fires, first aid and signalling for rescue.

In many ways, a land survival situation offers so many more chances than a sea survival situation. Speaking very generally, whereas on land the most serious threat to life will come from the cold or from the extreme heat, and the associated problems such as lack of water, in a ditching situation, this same threat is coupled with the permanent threat of drowning in the water surrounding you.

On land, too, you are less reliant on your own resources. Find yourself stranded in the vast majority of areas in the world and you can improvise with what nature has to offer you. In the sea you certainly do not get this chance due to a lack of natural resources.

This is not to say that being in a survival situation on land is easy; many of the conditions talked about will kill the average unprepared person in a matter of minutes.

Improvisation is the key, with an insight into how every little object could be useful for a dozen different tasks. Our lifestyles today have brought us so far from understanding nature that this skill is one that people have to learn quickly to adapt to.

Occupants of an aircraft can find themselves in a survival situation on land for one of many reasons. If an aircraft experiences any sort of mechanical problem serious enough that the crew have to attempt a forced landing; fuel exhaustion, possibly caused by setting an incorrect heading as was the case with the Varig Brazilian Airlines Boeing 737 that crash-landed in the jungle; an inflight fire or possibly through terrorism if there is a serious bomb threat or a hijacker on-board. For whatever reason, if the flight crew consider it necessary to make a forced landing and are unable to reach an airport, the occupants may find themselves in a survival situation. There is one further possibility and that is a Controlled Flight Into Terrain (CFIT), the increasing phenomenon whereby perfectly serviceable aircraft fly into a mountain, hillside or just the ground. Although this will rarely be a survivable accident, it has been known for occupants to survive such incidents. The Air Inter crash in France in which an Airbus A320 crashed into the hillside while on its final leg into Strasbourg, is an example of this. Eight people survived in the cold until rescue found the aircraft many hours later. Autopsies on the dead show that more survived the impact but died before rescue came.

The circumstances of a crash landing on land will vary according to the state of the aircraft and the surrounding terrain. A controlled landing due to fuel exhaustion, for example, will be much easier than a landing due to a major technical fault on the aircraft.

This book does not aim to cover the landing and evacuation in the event of an incident. All airline crew cover such eventualities in training and general aviation pilots should also be aware of their forced landing and evacuation procedures.

As in every emergency landing, the priority is to evacuate the occupants of the aircraft, as quickly and as safely as possible. Keeping everyone together and marshalling them away from the aircraft is then the next priority. If you have landed in a remote area, then the situation instantly becomes a survival one.

In a planned landing it should be clear where the aircraft is going to come down, at least in terms of whether the area is inhabited or not. If the area is remote, then preparation for the conditions that the occupants are going to be exposed to, should begin before the landing takes place.

This preparation is essential. To ensure a successful evacuation, no passengers can be allowed to carry anything with them off of the aircraft. Clothes that passengers are wearing when they evacuate, then, may be the only items that are salvaged from the wreck. Likewise, the crew should have a bag of items ready to take with them from the aircraft. Certain items should always be taken, for example, the survival kits, first aid equipment, water and signalling equipment. Anything else that can be put into this pack will increase the survival chances. This bag should be ready before the landing and stowed close to an exit so it can be removed from the aircraft after the passengers have safely evacuated.

If the crew have not carried out this preparation, they may be forced to leave the aircraft with nothing at all. It can not be assumed that the aircraft will be able to be re-entered. Fire may render this impossible.

The psychological impact often sets in once the evacuation has taken place. The initial fear of the landing itself and the relief of evacuating safely often gives way to panic once people realise the full extent of the situation that they are in, such as being miles from anywhere with little apparent hope of rescue.

This is when the crew role comes into its own. The crew have successfully evacuated all passengers from the aircraft; now they must take charge and keep everyone together. It is imperative that panic is not allowed to spread. Once several people are panicking, it will be very difficult to keep control and people may do things that put themselves and the whole survival effort, at risk.

Firm authority is what is needed. The passengers of any aircraft need to feel that the crew are in control. They need to trust them and that is why it is so vitally important that the crew of an aircraft know what they are doing in a survival situation.

Keep people together and get them to shelters as quickly as possible. Once in shelters, even if temporary, people will be reassured that something is being done to improve their plight. They will begin to feel slightly better too. They will be able to sit and rest and will be protected, at least in part, from the cold or the heat, or whatever the main threat is. They will also be centred around a small 'protected' area with a group of people. This will increase morale like nothing else.

As soon as you can, build a fire. This is vital in the cold climates and also in the hot climates at night, where it can get very cold. The fire will increase warmth and visibility. It provides a great boost to morale and will keep animals and insects away. It is also a great beacon that can signal where you are, should anyone be out there looking for you.

Attending to the injured is important but must be approached with an element of common sense. There will be those that can be helped and very probably those that cannot. As crew, you are responsible for the safety and welfare of all passengers. However, there is one basic rule to remember - the survival of the group is more important than the survival of the individual.

One of the best ways of keeping up morale, and improving people's survival chances, is to keep everyone busy. There will be a lot to do in any survival situation. Get passengers to help. There will be times when it is safer to sit quietly in shelters and do nothing, for example, during the day in the desert, but everyone can still feel that they have a role to play, even if they are not actively engaged in it at that moment. Being responsible for the survival effort, and especially your own survival, is one of the best psychological medicines available.

The aim in a survival situation is, not surprisingly, to survive. This means getting rescued as soon as possible and being alive to be taken home. Whatever the circumstances of the survival situation; whatever needs your attention, there is one thing that must be remembered regardless and that is keeping a watch for rescuers and being ready to signal.

Search and Rescue on land is covered at the end of the sections on land survival. Except in the jungle, where making yourself seen is incredibly difficult, you are again in an easier position on land than you would be in the sea. The main reason for this is that you are stationary on land. Once you have decided where to set up camp, you will stay there rather than drift away from the wreckage in the wind and current. Also, in a land survival situation, you may be able to remain close by the aircraft, once it is considered safe to do so, which will again help rescuers to find you.

As in all survival situations, follow the four basic principles of survival; Protection, Location, Water and Food. Be aware of the psychological impact of the situation that you are in. Control your own reactions before turning to help and control that of your fellow crewmembers and passengers and do everything you can to stem panic and enhance positive thought. Above all, be determined, adapt quickly, accept the situation that you are in and IMPROVISE. Believe that you will survive and you will.

12   The Desert

The Oxford English Dictionary defines a desert as an area which is 'uninhabited, desolate and abandoned'. It is defined geographically as an area that has very little moisture in the atmosphere. Although most people think of desert as sand dunes, the truth is that only very small parts of the earth's deserts are covered in sand. In the majority of areas the sand has blown away leaving a gravel surface.

For the purposes of this chapter, we will consider deserts as hot environments but it must be realised that, technically, many frozen regions are also termed as desert. For the purpose of this book, cold regions are covered in the Arctic chapter.

The lack of moisture in deserts means they are arid. Even the winds lose any moisture that was contained in them as they are heated up by proximity to the hot earth. Clouds are rare and so can never be relied upon to give any protection from the sun. Because of this lack of cloud, there is nothing to retain warmth at night and so day/night temperature extremes occur. In the day, the desert could quite easily reach in excess of 50 degrees Centigrade (120 degrees Fahrenheit); at night this can drop, frequently, to below freezing.

It is not uncommon for some desert regions to contain sparse vegetation. This will normally be in the form of cacti and similar water retaining plants.

All desert areas threaten the same dangers. The two quickest killers here will be the sun and a lack of water. Only those who have experienced the desert first hand will appreciate just how debilitating these two factors can be.

One fifth of the earth's surface is classed as desert and common airline routes overfly nearly all these regions. Passengers and crews within aircraft flying over any of these areas are in danger, if they encounter problems, of being in a desert survival situation.

Preparation before Landing

Think about the conditions to which you are about to be exposed. It will be hot; the sun will be a constant threat with next to no natural shade and there will be little, if any, water.

Water is obviously the first on the priority list of things to collect as without it, no-one will stay alive for very long. In the desert, a minimum of five litres of water per person per day is required to prevent the onset of dehydration. Collect up as much water as you have and take care not to waste any. Wrap full polythene water containers in something stronger to protect them against damage in an impact and while getting them out of the aircraft. Spilled water is wasted water.

Encourage everyone to drink before the emergency landing takes place. This will build up fluid content in the body. Avoid dehydrating fluids, though, such as tea, coffee and coke.

Other items that must be taken off the aircraft are the signalling and location aids as these may be your only chance of summoning rescue.

In addition to these you will need a full first aid kit and one of the most important items in this is sun cream. Light-weight clothing that covers as much of the body as possible, along with a hat, is also essential, to protect against the sun, insects and sharp vegetation.

Remember that the priority of a safe landing and evacuation take precedence over taking survival equipment from the aircraft. Although items to aid survival may make the difference between life and death in the desert, it is pointless getting them out of the aircraft if doing so is going to compromise the success of the evacuation.

Encourage passengers to put on as many items of clothing as possible before the emergency landing. Remember that it can get very cold at night so even thicker clothing will be useful. It is better to have them wearing what they need to take before hand so they can then evacuate without taking any belongings with them.

Remember that if the aircraft is not destroyed, it may be possible to go back on board at some stage after the landing, to collect more items.

Crew, however, should collect up anything that could be of use. Blankets, for example, can be used for the injured, for warmth at night, to use as shelters against the sun, and to use as a covering over rough ground. They could also be used, along with pillow cases and any other fabric, as make-shift clothes, shoes or hats. Improvisation is the key and you can find a use for almost everything.

After the Evacuation

The first priority will be to get everyone off the aircraft and to a safe distance. This will involve helping the injured, as in any evacuation.

The next priority is to find shelter. This is an urgent requirement as every second spent in the sun increases the dangers of dehydration, sun stroke and burns, all of which can lead very quickly to death. Shelters need not be elaborate and it is important to balance the benefits of such aids with the exertion needed in their construction.

As a general rule, all manual work is best done at night in the desert, when it is cooler and the dangers of the sun are not present. However, immediate shelters are essential if passengers and crew are to have any protection.

It is best not to move too far from the aircraft, unless close proximity of the wreckage is considered to be dangerous or unless help can be sought by travelling a short distance. Generally, aviation fuel will evaporate off in these temperatures quite quickly and so the aircraft will not pose a threat for long. It is best, therefore, to build shelters nearby because the aircraft wreck will be the thing most easily visible to rescuers and that is where they will start looking. This aid to your location will be essential as during the day, when the majority of searches take place, you will be sheltering out of the sun and so out of sight.

Use the shade around the aircraft as a sun shelter but do not try to live in the fuselage. In the heat it will be like an oven and during the cold nights, it will be like a fridge. Metal reacts very quickly to surrounding temperatures and is not a good shelter. The conditions will be so extreme that trying to stay in the aircraft could well prove fatal. The aircraft structure from the outside, however, can be ideal as a shelter from the sun because it will be big enough to offer shade to everyone on-board. Remember, though, that you will have to keep moving round the aircraft as the sun moves! Do not touch metal that is in the sun - it will be hot enough to burn and cause severe blisters.

Use the outside of the aircraft as a source of shade.

Do not try to reach help unless you are absolutely sure you can manage to do so. The heat is debilitating and will overcome the average person within just a few hours - especially if they are not acclimatised to this environment. A United States Air Force airman crashed his aircraft within clear sight of habitation in the desert. He began to walk and was found dead half way between the aircraft and his destination.

If you are sure that help can be reached, it is best to let just one or two people go to look for it. Moving everyone is just increasing the chances of fatalities.

The most effective action is to settle down quickly and simply, keep everyone together and keep still. The less exertion carried out, the better your chances of survival.

Immediate Shelter

Immediate protection can be found very simply. Do not worry about elaborate shelters until nightfall. Then you can afford the exertion to make them in order for everyone to keep warm. As an immediate measure, however, all you have to do is find shade.

The smallest amount of shade will make a big difference. The tiniest plants all give off a little protection and even if you can get only your head out of the sun, you will find it brings a little relief.

Look for natural aids in building simple shelters. Hollows in the ground or crevices, for example, are great in keeping everyone together - which makes everyone feel more secure and less likely to panic - and can be easily protected from the sun by using a blanket as a canopy. Weigh it down with rocks on the high points or tie it to bushes. This sort of shelter can be constructed in seconds and gives an area where everyone can sit and rest.

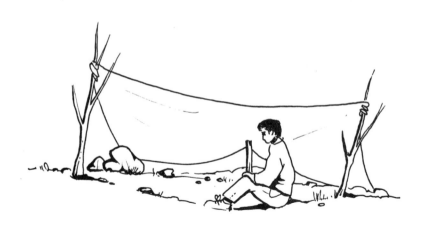

A simple shelter using a blanket propped on trees and bushes.

Using the outside of the aircraft as a shelter has already been covered. Parts of the aircraft, for example, the wings, could be used as the basic structure for a more permanent shelter later on.

A simple lean-to, such as a liferaft leaning against a pole or tree will also do as an immediate measure to get out of the sun. For this reason, it is important to try and retrieve as much of this sort of equipment from the aircraft as possible. Although designed for use in a sea survival situation, all equipment can have multiple uses.

A liferaft leaning against a tree can provide shade.

Be aware that just because you have erected some sort of canopy, you may not be keeping the sun out completely. It is important to keep the body protected as much as possible at all times and to administer suncream if you have it. Sunburn can become life-threatening if severe enough because it dehydrates and in this sort of sun, burning will be extreme.

A hat is probably the most useful item in protecting against sun stroke. If you do not have one, then improvise. The easiest to make, and probably far more effective than any hat, is the Arab head dress or *ghutra*. For this all you need is a large square of any cloth - the lighter in colour the better to reflect the sun - and some sort of cord which can be used to tie around the crown. This then protects the head, neck, ears and shoulders and, if big enough, offers protection over the face too. This shade over the face is a saviour if you do not have sunglasses.

Use this protection as it will help you. Keep your face shaded, it will reduce sweating and therefore water loss. When sitting still, wrap the cloth loosely over the mouth and nose and breathe through the nose to prevent moisture loss from exhalation. Talk only when necessary. All this may sound extreme when sitting in the office but learn from the Arabs: every little action like this will preserve life.

Another trick learnt from the Arabs; put a dampened cloth under the crown of your head dress to cool the head. Do not waste drinking water; urine will do just as well!

Making an Arab head dress is simple and could be a life-saver.

Shelter and water go hand in hand in the desert. By protecting the body and sheltering you are automatically helping to reduce dehydration. Covering the body fully not only protects against the dangers of the sun, but also slows the evaporation of sweat and so a loss of fluid. Wearing loose clothing traps an area of relatively still air between the body and clothing. This air also slows down the evaporation rate.

When sitting still in the shade, try to sit a few inches above the hot earth on a layer of rocks or vegetation. Even a couple of inches will make a difference in the temperature and this is something that you can do with little exertion. When building more substantial shelters, after the heat of the sun is gone for the day, digging below the surface of the earth is a good idea as here, too, you will find cooler surfaces.

If there is any surplus water that is not suitable for drinking - and you have no means of making it such - wet clothes with it. This will cool people and reduce sweating and therefore fluid loss.

Clothing is also important as a protection against sharp vegetation and insects in the desert. Everything that grows in the desert tends to be very sharp and there is a general rule that states: 'do not put the hand where the eye cannot see'.

On a slightly different tack, if there are members of the party who have lost their spectacles and who are reliant on them, give them a piece of card with minute holes in it. By looking though one of these tiny holes, the eye is forced to focus and so will improve vision. This will at least aid people to do essential tasks that they could not otherwise do because of poor sight.

Water

It may sound strange to put rest and shelter before water in the desert but finding water beyond the supplies retrieved from the aircraft is going to be a mammoth, if not impossible, task and not one to undertake in daylight.

While the body needs water desperately, every action undertaken will lose water from the body. Exertion for the sake of it in the desert is a killer and so if you are not going to improve your chance of survival by that exertion, do not do it. There is a rule in the desert survival situation and that is: 'Ration your Sweat, not your Water'.

Having said that, water is a number one priority. Without water, at temperatures approaching 50 degrees Centigrade (120 degrees Fahrenheit), you will die within two or three days - and that is if you carry out no exertion at all during this time. Debilitation to the point where a person is unable to carry out any useful task, such as attracting rescue, will occur well in advance of this!

Bringing water from the aircraft is an absolute priority. Even water that is not drinking water can be used in these desperate conditions. Water purifying tablets should always be in your survival kit and should be used in this case.

Use anything to carry water in but beware that you do not lose any. Plastic bags such as freezer bags are good but can split suddenly. Condoms are very good as they can carry a lot and are quite strong. Pour water into condoms slowly, however, to expand them to their full size. Just dipping them into the river will not do this. When full they should be able to carry 1.5 litres of water. Always protect supplies by wrapping the full bag in cloth or something similar as a double protection. Use whatever you have. Socks or stockings, for example, are great outer water carriers.

The water that you have brought with you may well be the only water that you can rely upon having. There are methods of acquiring water in the desert but they are labour intensive, take time to work and do not produce much for the effort involved. If there are just a couple of survivors, this may be sufficient but if there are two or three hundred passengers and crew, you are probably wasting your time.

It is as well to know the methods available, however, because trying to acquire water may be your last chance. You can distil urine and salt water, for example, and then drink it safely and so it is as well to know how this distilling process works and how to construct a simple still. Carrying out these tasks also acts in keeping people busy and in giving them a purpose to live. Both these factors are crucial to survival.

If you are lucky enough to find a lake in the desert, without an outlet, this has invariably developed into a salt water lake over the years. This water must be distilled or put through the reverse osmosis pump, before drinking. Distilling has the benefit here as you will be left with a residue of salt which can be consumed in very small quantities, preferably in water, to replace the salt lost through sweating.

Salt loss can become a serious problem and so replacing this commodity is important. The effects of salt loss include nausea, dizziness, muscle cramps and tiredness. Take salt either in this way or in the form of salt tablets, which should be contained in the survival kit.

Distilling

Distilling kits are often found in liferaft survival kits although as you need to boil the water, it is not very practical to carry out this procedure at sea. Even if you do not have a distillation kit, they can be easily improvised.

The water to be boiled needs to be put into a covered container that is positioned over the fire or heat source and a tube passed from this container into a sealed collecting tin. This collecting tin ideally needs to be placed inside another container that contains cold water which will cool the vapour of the water being boiled as it passes through the tube. Seal the gaps around the tube that comes out of the boiling container. This can be done with mud or wet sand, taking care that no sand particles fall into the water

A distillery.

Finding and Acquiring Water

There are ways of extracting water from many desert areas. Look for the signs that water may be present. Animal trails may lead to watering holes; bird flight may be a pointer to water; dried up stream beds or wadies may be a sign of underground water flow - the best place to dig is on the outside edges of the bends. Vegetation, too is a good sign that water is present. In dunes, look for water by digging at the base of the steep sides of the dune.

Any water in the atmosphere will condense in the form of dew in the early mornings. This dew will be found most easily on metal or rocks and can be soaked up with a cloth which can then be wrung out into a container or straight into the mouth. There will not be a lot of water collected by this method but every source must be tapped.

Another natural source, of course, is rain and although extremely rare in the desert, when it does eventually rain, it rains hard. Survivors must collect up every drop while they have the chance and store it in a sheltered place so it does not evaporate off when the sun shines down again.

Transpiration Bag

Exploiting a plant's natural condensation will give small amounts of water. All plants draw water from the roots and distribute it to the leaves. The greener the plant used, the more water produced. The technique is simple. Cover the foliage of the plant with a plastic bag and seal this bag tight at the bottom of the plant. Dig a small depression near the plant and push the plastic into it to give a collection point for water produced as the plant transpires.

A transpiration bag.

As an alternative to this method of water procurement, handfuls of fresh foliage can be used instead of a growing plant in exactly the same way.

Solar Still

By using a solar still you are almost guaranteed a supply of water, albeit possibly very small.

To make a solar still, dig a hole in the ground that is about one metre across and three quarters of a metre deep at the central point. Put a water collecting tin in the hole at this lowest point and fasten a long piece of tubing into the container. Bring the other end of the tube out of the hole and lay it to one side on the ground. Put some broken vegetation into the hole around the collecting container to increase water output. Urinating into the hole, but avoiding the collecting tin, will also increase water production.

Now spread a large sheet of plastic over the entire hole, bringing the tube out from under it, and seal the plastic all round the edge with rocks and sand. The centre of the sheet should also be weighed down by a stone so that it is depressed by about 14 inches.

The still works by the sun warming the ground under the sheet and evaporating any water present. The air under the sheet therefore becomes saturated with this evaporated moisture, which eventually begins to condense on the underside of the plastic before running down and dripping into the collecting container.

For this device to work, the plastic sheet must be sealed around the hole in the ground; the depressed sheet must be clear of the water container or water will be lost on the outside of the container and the sheet must not touch the sides of the hole at any point or water will seep back into the soil rather than run down into the container. A clear plastic sheet is the most effective for this task; especially with the underside slightly roughened. Remember that if you do not have a ready made container, make one. Aluminium foil, for example, can easily be shaped into a container. Fold and shape several layers of foil as required and stiffen by folding the corners for stability and support.

A solar still can be expected to produce up to two litres of water in 24 hours, especially if this production is encouraged by the presence of vegetation and urine in the hole. From this it can be realised that one still will not provide water for many people and that a great deal of these will have to be dug if large numbers of survivors are relying on this water as their only source. Remember, too, that over time, each still will reduce in yield and will have to be re-sited.

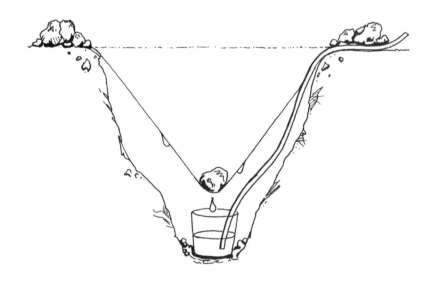

A solar still.

Filtering and Purifying

Any water that is extracted from the streams or from under the ground must be purified before drinking. Water may be the first priority but vomiting and diarrhoea from illness caused by unclean water will dehydrate far more than the lack of water will.

Water filters can be obtained and kept in the survival kit. If you do not have one of these, improvise. As a basic measure, use a sock or piece of cloth for the first stage. Filtering the water through this, being careful not to waste any, will get rid of large particles; for example, all but the tiniest creatures, leaves, mud particles and other foreign objects.

After this, water must be purified so that it is safe to drink. Purification tablets are the safest method. Make sure they are the type that can be used to purify even the dirtiest water. These are usually iodine or potassium tablets. Boiling is another method of purifying. Make sure the water boils thoroughly and evenly but do not boil to the point where water is evaporated off and lost. Boiling for five minutes will ensure safe water. A combination of these two last methods will ensure completely safe drinking water. Boil first and then add purification tablets.

A reverse osmosis kit, that should be supplied in every survival kit, is also a means of acquiring safe water. Make sure that this item is retrieved from the aircraft if at all possible. Because its obvious use is for a ditching and sea survival situation, it may be stored in the liferaft kit. Any water is safe once pumped through the reverse osmosis kit except that which contains chlorine.

Water gained from transpiration bags or solar stills will be purified and safe to drink.

Plants as a Water Source

There are one or two desert plants and cacti that can be used as water sources but as vegetation is different in every geographical area, it would be unwise to guess which plants are safe to use. There is a method of testing plants to see whether they are safe to eat; this will be covered under the food section; but as a rule, only use plants as a water source if someone in the group knows the vegetation in that area. You should definitely not drink any milky or coloured plant juices as they are likely to be poisonous.

Water Rations

The most important thing to remember about all rations is equal division. This is essential for good morale and also to prevent panicky passengers from rebelling and compromising the whole survival effort.

Use common sense as to water distribution according to the amount of water available. Enough water needs to be drunk to prevent serious dehydration. If there is ample water, then everyone should drink five litres a day but this will rarely be the case.

Drink planned amounts at planned times and stick to these so everyone has a routine that they can adhere to. Monitor the amount of urine being passed by each person each day. The amount drunk should exceed the amount passed;

ideally by half a litre but at least by a small amount to replace fluid lost from perspiration and transpiration. By doing this, the amount lost is at least being replaced. Dehydration will begin but the worst effects of it should be kept at bay, at least for a while.

Even if supplies are relatively good, this amount should be stuck to in order to preserve supplies for later on. Do not assume that rescue will be instantaneous. If signs of dehydration begin to show (see Chapter 16, First Aid on Land), increase the amount little by little.

Drink in the evenings or early mornings rather than in the heat of the day. Sip water and swill it around the mouth, which will be very dry. Do not gulp it down quickly. This will make a dehydrated person vomit which only loses more fluid. To allay thirst in between drinking, suck small stones or chew grass and string.

Do not eat unless you have ample water. Huge amounts of water are needed for digestion and food is not essential to survival for the first few days. Do not drink alcohol.

Constipation is common in these conditions; it is the body's way of preserving fluid. Urine, too, will be very thick and yellow.

More Substantial Shelters

Once the heat of the day has passed and there is no sign of rescue, it is a good idea for survivors to begin to build slightly more substantial shelters. These will give better protection, both throughout the cold nights and during the hot days. A good shelter will also give a feeling of security and will allow some rest and sleep.

Even now, shelters do not have to be very complex. You may be working through the cool nights but you must still be aware that exertion saps fluid and energy from the body.

Use any natural formations available to you as an aid such as caves, trenches or gullies. The shelter needs to be able to protect you from the cold at night and, probably more importantly, from the sun during the day so choose the naturally most damp and shady areas, if there are any. Be aware, though, that if you find such cool places, you can be sure that all the undesirable creatures of the desert will already be living there and so you will have to evict them before moving in.

Try to dig down about 18 inches; here the soil will be cooler. If this proves too difficult, use rocks and foliage to raise the bed above the ground. This, too, will insulate against the extreme heat of the earth. Extra foliage can be used to protect from the cold at night, if necessary.

Be aware what sorts of plant are used in shelter making. As a general rule, if there is any growing vegetation in the desert, it will be prickly and very sharp. Even the berries will be covered in tiny thorns which must be removed before eating. In some areas, though, it is possible to find dried grass, tumble-weed and other types of vegetation.

Protection from the day time sun is the main point of shelters and so this shelter must be thick enough to have a positive effect. Keep shelters simple, though. The diagram below gives an example of a simply made but efficient shelter. The cover could be constructed from any cloth, for example, blankets.

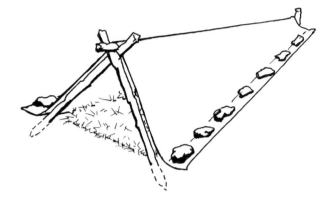

A simply constructed shelter.

As an even simpler shelter, string chord through the apex of the blanket and tie each end to a tree or bush, again weighting down the edges with rocks and stones. This removes the need to lash poles together.

Be aware that the covering used may not prevent all the sun's rays from coming through. Two layers are really needed to do this. If equipment is sparse, keep the body covered as completely as possible as the second layer of protection and never forget the head.

Remember that just about anything can be used in the survival effort. Excess clothing, blankets, pillow cases and even seat upholstery can be removed from the aircraft and used as protection against the sun. Sew small bits of fabric together if required. Some cacti have sharp points attached to strong cord that make these plants excellent for use as sewing tools. Simply strip the green foliage away to expose the cord and use the sharp point as a needle. When lashing poles or bits of wood together, use a diagonal lash around poles being fixed together at right angles and a shear lash for those being tied in parallel.

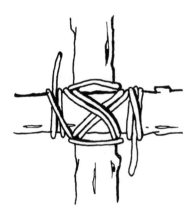

An example of lashing.

In terms of cord for these tasks, again use anything. Many survivors will tell you that one of the lightest weight and strongest substances is dental floss.

If you are lucky enough to find a cave, many of your prayers are answered, especially if it is big enough for every one to shelter in. A cave will keep cool in the day but will become very cold at night. Light a fire towards the back of the cave for warmth and put a barrier at the cave's mouth to keep the heat in. The smoke will rise and follow the contours of the cave ceiling, keeping fresher air lower down. Make sure, though, that there is room at the mouth for the smoke to escape. A fire lit at the mouth of the cave is unlikely to lose its smoke outwards. It will probably be blown in.

This is a perfect shelter, as long as you can keep a fire burning. (This will be covered in Chapter 15.) It keeps everyone together and out of the wind and sun and with a fire will be a deterrent for beasts. Remember though, to keep a close look out for rescue and be ready with your signalling at the first sign of help. The inside of a cave cannot be seen from the air and without the hustle and bustle of an outside camp, you could very easily be missed by those searching the area.

A cave can make an excellent shelter in the desert. Make sure that there is adequate ventilation at the entrance for smoke from a fire to escape.

Food

Food is not considered a priority by this book because it is not essential to the survival effort, especially in the desert. Rescue is almost certain to come within three days, if not sooner, and a person can live this long without food. In addition, as vast quantities of water are required to digest anything, eating will actually confound the survival effort unless you have copious quantities of water.

In this unlikely event, and remember that the ideal amount required by each person in the desert, in order to stave off dehydration, is a minimum of five litres each day - without taking into account water used for digesting food - you will need a huge amount of water to be able to consider eating.

If, for any reason, eating food is considered, it is important to make sure that any food eaten is safe. Food in the survival kits and any taken off the aircraft will obviously be okay to eat, as long as perishables are not used. Remember, though, that safe storage for food is essential because there are many insects and reptiles in the desert who may themselves fancy a snack.

If food is sought from the desert itself, care must be taken to ensure that this food is safe for consumption. There is a recognised method of testing food - and this can also be used for testing potential water supplies that are found within plants.

There are two sources for food in the desert; the first is vegetation and the second is animals. In the very rare event that you wish to locate food, it is far safer and easier to stick to the vegetative variety. You then do not have to get into acquiring skills such as snaring, killing and preparing dead animals. You also do not have to run the risk of having to fight with the possibly dangerous animal when trying to kill it.

In addition to these factors, vegetation takes less water to digest and is generally easier for a weakened digestive system to cope with.

Testing Food

Only allow one person to test each new variety of plant, just in case. Carry out the whole test and never assume that the plant is safe before finishing the entire procedure. If in doubt, do not eat the plant. Never assume that because an animal has eaten the plant it will be safe. It may not be suitable for humans. Illness caused by poisonous plants will at best increase dehydration and at worst, kill those who may otherwise have survived.

First inspect the plant that you wish to eat or extract water from. If it is slimy or worm-eaten, leave it alone. This means the plant is old and past its best. The plant will have very little food value and may even have become toxic. With these plants, the worms will probably have more edible value than the leaves themselves.

Smell can also give you a clue. Crunch a small portion of the plant and smell it. If it smells of bitter, discard it.

After these preliminaries, the first step to testing is to squeeze a little of the plant's sap onto a sensitive part of the body, for example the skin on the inside of the wrist, and leave it there. Wait a few minutes and see whether an adverse reaction occurs. If you experience irritation, rash or swelling, discard the plant immediately.

The next stage is to try this same procedure out inside the mouth. Only if you have had no reaction to the presence of the sap on the skin, squeeze a little sap first onto the lips. Wait a few minutes and then do the same thing in the mouth and finally on the tongue. If there is still no reaction, try chewing a small piece of the plant but do not swallow it at this stage. Warning signs to look out for are burning or irritation and swelling.

If you experience none of these adverse reactions, swallow a piece of the plant. It is imperative, however, that you do not eat or drink anything else after swallowing the plant, for at least five hours or the true effects of the plant can not be assessed. If there are no reactions after this time, for example, stomach cramps or violent pains in the lower abdomen, nausea and vomiting, the plant may be considered safe.

Even then, though, it is important that the plant is introduced to the diet slowly. The stomach will not be used to such a thing and may reject it if it is faced with too much of it.

Eat little bits of the plant over time and the stomach will have a chance to adapt. Also, eat the plant as you tested it, ie cooked or uncooked. Some plants may be safe when cooked but poisonous raw and others may turn poisonous when cooked.

Plants to Avoid

Always avoid bitter smelling or tasting plants; those with milky sap and those with barbs on the leaves and stems. Most desert plants, however, will have tiny thorns, even on the edible fruits, which must be removed before eating.

Lavatory Arrangements

Quite simply, choose an area for depositing all waste and stick to this plan. Make sure this area is down-wind of your camp and well away from any potential water sources. Burying waste is the most hygienic way of dealing with it. Try to encourage people relieve themselves during the cooler night hours so they do not have to get out of their shelters during the hot days.

13   The Jungle

Incident

On September 3, 1989, a Varig Brazilian Airlines Boeing 737-200 made a crash landing in the jungle.

At the beginning of the last sector of the flight, the flight crew selected course 270 instead of course 027 - the correct course. This took the aircraft right out over the Amazon Jungle, where there are few aeronautical aids. As a consequence of the incorrect course being set, the aircraft ran out of fuel over the jungle, miles from anywhere.

The Captain informed the Purser that a forced landing would have to be made. The passengers were informed and all cabin baggage was stored in the lavatories. Cabin crew re-located passengers closer to the emergency exits and handed out spare seat cushions for passengers to use as extra head protection.

As the left engine failed, the Captain instructed the passengers to brace. Shortly afterwards, the right engine failed and the flight recorders failed. The aircraft was descending in the dark, with no visual references. The landing gear was up.

The trees, over 100 feet in height, ripped the aircraft wings apart. The impact was intense and twisted the cabin floor. All seats, with the exception of two rows on the right and one row on the left, all at the very back, came loose and piled up at the front of the cabin. The overwing exits and doors 1L and 2R were jammed. The escape slide at door 1R was punctured by tree branches and failed to inflate.

There were 54 occupants on the flight. 12 died in the incident, 17 were seriously injured and 25 slightly injured.

The incident happened at night and visibility in the jungle was very poor. Little was done to free those trapped among the piled up seats during that first night because no-one could see. Parts of the overhead bin compartments had also collapsed and were blocking areas of the cabin.

The Captain broke the cabin windows to let some air into the aircraft. The Purser improvised with a straw, offering water throughout the night to a small child that was trapped in between seats.

The following morning, the task of getting everyone out the aircraft began. Immediately, the shortage of equipment became apparent; in particular water, first aid kit and tools. The Captain took charge, aided by the Purser and organised everyone into helping with the survival tasks. One survivor had a specialist knowledge of the jungle and was able to track water for the survivors.

The Captain also activated the Emergency Location Beacon, which, needing water to activate its batteries, had to be surrounded with water from melting ice and urine.

A Rescue Co-ordination centre received the beacon's signals and assigned three aircraft to search for the aircraft. The survivors heard the aircraft engines and lit fires to try and draw attention to their exact position. The survivors had not taken into account the very thick vegetation cover in the jungle, however, which will not allow anything to be seen through it and so the smoke was wasted. The smoke from the fire did, however, help in keeping the mosquitoes away but many of the injured reported that it disturbed them.

The aircraft searched all day but could not find the survivors. Darkness fell yet again. A second night in the jungle and both morale and hope were fading.

The next day, a group from the crashed aircraft set out to search for help and did manage to reach a plantation farm, from where they could radio their position to the authorities. Search and Rescue aircraft continued their search and at 16.34, visual contact was eventually made with the survivors.

Early that evening, an injured woman was winched to safety but died on the way to hospital. The other survivors had to wait a few more hours while experts, who were flown in, opened up enough space in the thick jungle to allow helicopters to land. Everyone was then rescued, some 44 hours after the crash landing took place.

* * * *

The term jungle is used to describe those remote tropical areas of the earth's surface. It is the most difficult climatic zone to describe in detail as these regions include areas of desert, swamp, and forest. Typically, though, the jungle is made up of vast areas of rain forest which, as its name suggests, experiences huge amounts of rain.

The climate within these regions varies very little although there is a propensity for violent storms at the end of the summer months. Except at the highest altitudes, the temperatures are steadily high, although not at the extremes found in the desert regions; usually around 37 degrees Centigrade (98 degrees Fahrenheit) although it will cool a little at night.

The rain, when it falls, falls hard and relentlessly and can lower the temperature slightly but when it stops, the temperature soars again very quickly. These areas are prone to flooding. Despite the temperatures being lower than in many desert regions, do not be fooled into thinking that life will be any easier. The humidity in the jungle is at 90 per cent most of the time and this is extremely tiring and debilitating. So severe is the humidity that jungle regions are often wreathed in a mist.

The picture being created, then, is of a hot, steamy and rainy place. The tree cover is extreme, with trees growing to 200 feet in height and with smaller trees extending the canopy so thickly that barely any light penetrates through to ground level.

The environment many sound more conducive to survival than either the desert or Arctic regions and in many ways it probably is. There is a plentiful water supply, if you know where to look for it; the extremes of temperature are not present and the burning sun is not a factor. True - but there are also many things about the jungle that make the desert and Arctic regions seem like a picnic.

Water may be plentiful in the jungle - but it is far from clean. Germs, disease, parasites, unsavoury creatures such as rather large snakes, scorpions and bugs of every shape and size are also rife! This is the home of the mosquito, the leech, the poisonous caterpillar and an assortment of bees, wasps and hornets, to name but a few. The ground will be crawling with these creatures. Depending on where you are in the world, you may also be sharing your new territory with larger animals such as wild pigs, tigers and other large cats, many of which would love to eat you.

On top of this unpleasantness, probably the most serious consequence of crash landing in the jungle is the fact that rescue is so difficult because from the air it is completely impossible to see anything through the canopy of vegetation. From past incidents, it has been shown that even a thick swathe cut through the trees by the crashing aircraft itself is not readily visible. The chapter on Search and Rescue covers this subject in more detail but rescue can take time in this environment.

Preparing for the Landing

In view of the fact that rescue can be so difficult, it is especially important that the flight crew give the position of the aircraft as accurately as possible before the landing. This will not make rescue any easier in terms of being seen by the rescue crews but it will at least give a smaller search area for them to start with. Also, if rescue takes place from the ground, the rescuers should be able to pinpoint your position reasonably accurately.

In terms of what to take from the aircraft, the usual rule of 'as much as possible' applies. Remember that you can always retrieve more once the aircraft is on the ground, presuming that it has not been destroyed. Adequate clothing to protect against the 'wild life' is probably one of the most important things to take. This should be donned before the landing so people can evacuate wearing what they need and not having to carry anything.

Water is also important. There is usually a plentiful supply in the jungle but any clean water that you can take with you will prevent you from having to collect and sterilise water before it can be drunk. It should not be assumed, however, that water will always be easily available in the jungle. Although plentiful, it may require some searching for. When the Varig Boeing 737 crashed in the Brazilian jungle, for example, it took a passenger who knew the jungle environment very well to lead the survivors to the water sources. These were not evident to the average passenger and crewmember.

First aid kits are essential and should contain insect repellent and water purifying tablets.

Any sheets, blankets or excess clothing are also useful for making shelters, covering the ground you are to sit on, protecting items from insects and for wrapping up supplies and storing them safely.

In the jungle you need to be as well covered as possible to protect against scratches and insects. Clothing will be soaked through with the humidity and with perspiration from the body within seconds but it should not be removed as this protection is needed at all times. Drying clothing is not really an option anyway with this much moisture in the air unless you can build a fire.

Initial Protection

Despite any unpleasantness, the initial actions in the jungle are not so critical as other environments in which you may find yourself. The immediate threats such as the relentless sun, the freezing cold and the drowning sea, all of which can kill very quickly, are not present here.

Evacuating the aircraft safely is obviously the first priority. Presuming the area surrounding the wreckage is safe after everything has settled down, ie the engines have cooled and the fuel has evaporated, it is probably as well to stay close by. Although the wreckage will not be very visible to rescuers, there is no real point in moving away from it unless it poses a threat; you are sure you can reach help easily or you can move to an area where there is a clearing in the trees which will enable you to make yourself visible to rescuers.

Also, as another argument not to move from the aircraft, this may be the only time in a survival situation when you want to use the aircraft as a shelter. There is bound to be major structural damage to the aircraft when landing in the jungle, because the aircraft will almost certainly be crash-landing among trees. Even so, if it is considered safe to do so, it may provide a good shelter. Although it will be hot, direct sunlight will not be burning down on the wreckage thus turning it into an oven. The wreckage will protect from the rain and also, to a small extent, from the creatures and insects of the jungle.

Keep everyone together and give them tasks where necessary. Do not let people wander off on their own, unless they have a specific knowledge of the environment and think they can find help or water or something else that will improve the survival effort. It is very easy to get lost in this environment and if people disperse, it could lead to panic and a general drop in morale. As in any survival situation, it is up to the most senior crewmember there to take command and delegate through the rest of the crew.

Protection against Insects

One of the first priorities is for each individual to cover their body. This should be done on the aircraft before the landing if possible. Keep everything covered, unless you want to share part of your anatomy with a host of creatures. As the body perspires, for example, it gives off salt and that is exactly what attracts these creatures. If they had the chance they would happily take up residence in armpits and groins. Try to tuck trousers into socks or wrap fabric around the outside of trouser bottoms to seal them and tuck scarves or handkerchiefs into the neck.

If, for any reason, there is a shortage of shoes and clothing, improvise and wrap anything around your body. Do not, for example, let anyone tread outside without something on their feet.

Try to cover the head and make a fringe with cloth or vegetation, that hangs down over the face, to protect against insects and mosquitoes. This is especially important at dawn and dusk. Netting is the obvious material but as you are unlikely to have any, use a piece of clothing, for example, a tee-shirt.

It is important to keep covered during the night as well though. Use strong insect repellent if you have any. Oil, fat and mud can also be used on the hands and face. Smoking cigarettes or lighting a fire helps keep insects at bay.

Brush off caterpillars and similar creatures as quickly as possible as some species can be poisonous. Brush in the direction in which they are going or their spines will stay in your skin and will cause nasty irritations. Also remove leeches from the body. Although not dangerous, they are unpleasant. Their bite does not hurt and they will generally drop off when they have had their fill of your blood but they do inject an anticoagulant which means that you will bleed for some time. Remove them by burning them off with a cigarette end or by dabbing them with salt or alcohol. Do not pull them off. The jaws tend to stay in your skin!

Do not leave uninhabited clothing and shoes on the ground and always shake them out before putting them on. If you have been resting, or asleep, move with care when you rouse. There are bound to be one or two creatures that have used you as a shelter. Be careful when putting hands into bags and pockets at all times.

Water

Water is essential in the jungle. Do not be fooled into thinking that it is not a priority because it is a geographically wet place. The heat and humidity will cause people to perspire every bit as much, if not more, than in the desert and that water must be replaced.

Water brought from the aircraft is the best; it is clean and readily available. It is unlikely, though, even if you are able to successfully retrieve water from the wreckage, that you will be able to bring enough water to satisfy demand. The ideal is for every person to drink a minimum of five litres a day. This was the figure in the desert and the same applies in the jungle. The only difference is that this should be a more realistic target in the jungle because water should be more readily available.

Water from the aircraft will keep dehydration at bay while more water is located, though. In the jungle, it is not as important to restrict activity to the night time because although it will be slightly cooler at night, you will not have the dangers of the sun during the day. Visibility at night, too, will be very poor.

Remember that the vast majority of water found in the jungle will not be clean. This is important as it will probably be the source of every tropical disease known to man. Do not fall into the trap, therefore, of drinking it before doing everything in your power to sterilise it. The exception to the clean water rule is water obtained from plants such as the water vine.

Much the same techniques apply to finding and sterilising water as in the desert. Look in the lower lying areas, especially where there is much vegetation. It may be necessary to dig for the water but it should not be too far below the surface. Look for animal tracks that may lead to water sources such as springs and pools.

Strain the water first and then boil it or add purification tablets or, preferably, both. The use of distillation kits or reverse osmosis pumps will also work.

One very useful source of water in the jungle, of course, is rain water and this will not need to be sterilised as long as it is collected on the way down. Once it has settled on the ground or in pools it must then be treated like any other water source found in the jungle.

Be suspicious of pools or lakes that may contain natural gases or chemicals. Large bubbles coming regularly to the surface of the water are likely to indicate gases. Bubbles from an animal will move over time, as the animal moves away.

Pools with no vegetation around them, or as an even greater clue, with signs of dead animals nearby, may contain toxic chemicals or vapours. Try not to drink this water; if it is your only supply then take extreme measures to ensure that it is sterilised.

Food

Food is not a priority in any survival situation as you are almost certain to be rescued within, at most, three days. If food is available and there is a very good supply of water then eat it but it is not a calamity if you eat nothing. It is more important to preserve fluid limits in the body. Avoid eating food that is likely to make you ill as this will have an adverse effect on fluid limits.

The methods and advice on selecting and testing food can be found in the desert survival chapter. Although there will be a plentiful supply of both vegetation and animals, it is best to stick to vegetation, if you wish to eat anything at all. If possible, and you have the necessary water supplies, stick to the contents of the survival kits or to what you can retrieve from the aircraft.

Shelters

You should have no shortage of shelter building material available to you in the jungle. The best piece of advice is to build shelters and sleep off the ground in order to give best protection against insects on the jungle floor. A hammock would be ideal.

If this is not possible, at least use a thick ground covering. Bamboo, bark, branches and leaves can be used for this task. Then, if you have one, use the luxury of a blanket to cover this vegetation. You do not have to worry too much about protection from the extreme cold or the sun in the jungle so the task of building shelters is not as critical in this sense. You will need to shelter from the rain, though and it does become quite cold, in comparison to day time temperatures, at night so prepare for this.

Creatures are probably your biggest threat and as in the desert, you are going to have to evict them when you want to build your shelter and again every time you wish to return to it.

Use the aircraft as a shelter, or bring items from the aircraft. Life rafts and evacuation slides, for example, will provide something to shelter under or something to sit on to keep you above the ground. Watch out for sharp objects and vegetation that may cause damage to rafts.

It is not beyond the realms of possibility to remove any loose seats from the aircraft. Any seats where the fittings have failed during the impact can easily be used for sitting on to keep people off the ground.

Building shelters should not be too difficult in this environment. Thin branches with thick vegetation can be bent down, for example, to provide a good covering that will keep much of the rainfall off.

If you are more adventurous, you can make a simple lattice of sticks by lashing them together, covering it in thick leaves. This too will provide protection against the worst of the rain.

Bent down branches can provide shelter.

Choose where you build your shelters, within a small area around the aircraft or wherever you have set up camp. Choose the driest areas; usually on gentle slopes. Avoid valley floors which tend to be wet and higher slopes as they tend to be exposed. Look out for animal tracks that may suggest a highway to the watering hole that goes straight through your shelter!

If you are unfortunate enough to crash land at higher altitudes, ie higher up a mountain, then you will have to protect against the cold wind as well as the rain. Moving down to a slightly lower altitude, where you can find trees to use for shelter, may be your best chance of survival in this situation.

Man-eating Animals

It is possible, depending on where in the world you crash, to be in an area that is home to large animals, such as tigers. There is little you can do to protect against a tiger if it is hungry. Fire is the best deterrent, apart from making sure that everyone stays together so there are no individuals at the extremities of the camp. The aircraft fuselage may provide some protection in this situation, especially if there is a fire at the entrances, to deter animals. Be aware of flammable substances close to the aircraft, however, before lighting a fire.

Lavatory Arrangements

As in the desert, make sure that all waste is deposited well away from the shelters and, most importantly, water sources. Great care must be taken when going to the toilet in these conditions, to avoid being bitten or stung in nasty places!

14 The Arctic

Arctic, or Polar regions, are, as their names suggest, found at the North and South Poles. Generally, Polar regions are considered to be any area of land that is further north, or further south, than 60 degrees 33 minutes. It is in these regions that a knowledge of Arctic survival becomes essential although it must be realised that at very high altitudes anywhere in the world, these skills could also be a life saver.

Arctic conditions in areas regularly crossed by passenger aircraft present themselves in the northern reaches of Alaska, Canada, Scandinavia and Russia. With the increasing adoption of FANS/ATM and the relaxation of Russian air space restrictions, more and more commercial aircraft are flying over the North Pole. Islands such as Iceland and Greenland are also counted and it must be remembered that in winter, severe conditions could occur far further South. What is termed as Arctic conditions for meteorological purposes could be very different from a climatic condition that would still kill you.

In addition to these Arctic regions, mountain areas also present great risk. High altitudes in the world's hottest climatic areas will also present sub-zero temperatures, wind-chill factors, snow and ice. Such regions include the Rocky Mountains, Andes and the European Alps.

Arctic regions are transversed by aircraft on a daily basis. Think about it. Any aircraft that flies from Europe to America, for example, flies over Arctic regions. Any aircraft that flies over mountain regions, too, is at risk, in the event of a problem, of being in an Arctic survival situation.

The problem of Arctic survival, however, can be brought far closer to home. People die of exposure in conditions and temperatures that would not be considered typical of an Arctic region. People die when walking out on moderately sized hills in the UK, dressed for the activity they are undertaking. The Air Florida crash, in which a Boeing 737 went into the Potomac River just after take-off, left survivors clinging to wreckage in frozen water. This was not an Arctic region; it was America in winter. The moral of this? Do not underestimate the importance of Arctic survival.

Returning to the two main Arctic regions, let us look at the conditions to be expected in each. The South Pole is land locked, covered by a sheet of ice. The North Pole, however, is largely formed of ice that floats on the sea. At the North Pole, the winters are long. The days are almost totally dark in winter, contrasting with 24 hours of sunshine in the summer. Temperatures in winter fall to -56 degrees Centigrade (-80 degrees Fahrenheit) and never climb above freezing. In Siberia, a temperature has been recorded of -69 degrees Centigrade (-94 degrees

Fahrenheit). Summer temperatures can climb to 18 degrees (65 degrees Fahrenheit) and much higher in land areas, such as Russia. In the summer months, blood-sucking insects are prevalent.

The temperatures in the South Pole region tend to be lower than at the North pole and the winds are very strong; 110 mph is not uncommon. This causes an unthinkable wind chill effect and blizzards.

The biggest danger facing the occupants of a crashed aircraft in this region of the world is the cold. It will be extreme and hypothermia will set in quickly. All efforts must go towards protecting everyone from the cold, the wet and the wind as this is the only hope of survival. Shelter must be found fast.

Preparation

Putting on extra clothing is the most useful thing people can do when anticipating a crash-landing in an Arctic region. General aviation pilots who are flying over such regions have no excuse not to be suitably dressed at all times when in the aircraft. Airline occupants will certainly not be. Imagine leaving the UK in the Summer for Florida - no-one will have even UK winter clothes with them, let alone anything more suitable for the Arctic regions.

Any items of clothing that can be found must be used. Improvise. Flesh will freeze in seconds so it is important to cover every surface of the body, including as much of the face as possible. Remember the head, hands, feet and ears. Wool is one of the best fabrics to wear. It does not absorb water and is warm, even when damp. Cotton, on the other hand, does absorb water and the wearer will lose an enormous amount of body heat when it is wet. Layers are vital as they trap air between the garments, forming a natural insulation.

Ideally, outer garments must be woven closely enough that they are windproof but not unbreathable. Many waterproof garments do not breathe at all, for example, and will not allow water vapour to escape, creating condensation on the inside. Condensation is moisture and any moisture reduces the insulating quality of fabric and air spaces. Wet clothes reduce insulation by 10%. It must therefore be your aim to keep as dry as possible.

The atmosphere is generally very dry in most Arctic regions. You are, in fact, in a cold desert and so there is very little moisture in the air. Waterproof clothing is not, therefore, a necessity because any snow that falls on you will be very dry and will not seep into your clothes. The only time you are likely to get wet is if you come into contact with water, or with snow for extended periods of time, ie lying down on it or if you have snow on you that is then taken into a shelter where it will warm up and melt.

It is vital that heat is allowed to build up inside clothing. This provides an essential warmth that may make the difference between life and death. Any gaps in clothing will allow this heat build up to escape. Make sure, then, that clothing is sealed. A draw-string around a hood, if you have one, is very important, as is tying something around the neck, sleeves and ankles to prevent gaps. Improvise - anything can be used for this task. Cloth can be tied around the head and neck if you have nothing else.

One exception to this rule is when undertaking physical activity. Here, the body should be allowed to ventilate to prevent overheating and a build-up of moisture.

Inflated lifejackets can provide very good insulation, especially if worn under a coat or jacket. Although bulky and in many ways impractical, the air in the chambers will warm and will insulate effectively.

Make sure your feet are well protected. Feet tend to suffer more than other parts of the body because they are close to the ground where shoes and boots are often wet and they perspire a lot. Again, wrap your feet in layers and make sure garments are not too tight because this reduces insulation.

In thick snow, it will probably be necessary to wear snow shoes. Remember that if feet sink deep into snow, you increase the chances of getting wet and this increases the dangers of cold.

Snow shoes are easy to improvise. All you need is something lightweight that has a large surface area. Tie sticks together in a sort of tennis racket shape or use evergreen boughs or parts of the aircraft fuselage. The larger the surface area of the shoes, the more efficient they will be.

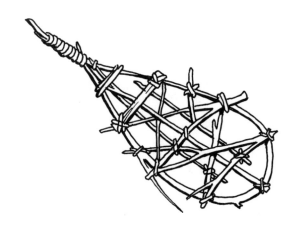

Snow shoes.

Take sun glasses, if you have them, to reduce the effects of snow blindness. If sunglasses are not available, protect the eyes by covering the face in strips of fabric or by making a slit mask. This is a covering for the eyes with very small slits in it for you to see through. This will reduce glare.

It is also important to note that a slit mask, with very small holes in it, is a useful tool for anyone who has lost their glasses and who is unable to do much without them due to poor eyesight. Looking through tiny holes forces the eyes to focus much more clearly than they otherwise would and really does improve vision. (This is covered in more detail in the Desert survival chapter, 12.)

Another useful hint for reducing the glare of snow blindness is to blacken the surrounds of the eye and cheeks with charcoal, or a similar substance.

Put as much clothing on inside the aircraft as possible so evacuation is not impeded by trying to get things out of the aircraft. As always, you can re-enter the aircraft once it is on the ground, if it is not totally destroyed.

Other items to take include fire lighting materials as this task will be crucial to survival. Fires will enable you to melt snow and ice for water, keep away wild animals, keep warm, dry clothes and signal.

First aid kits are also important as is anything insulating and anything that could be used in building shelters. Take water, if you have it, although it will freeze and it will be possible to acquire water from snow and ice. Signalling equipment is also a vital commodity. This is the only survival situation in which it is worth taking food. Usually food causes too much trouble because of the lack of water but in the extreme cold, if you can acquire enough water, then simple food, such as glucose tablets and the survival bars found in survival kits will help replace energy and warmth that may be a lifesaver. Providing you have the ability to light fires, obtaining water will be no problem in these regions.

After the Evacuation

The one first priority is shelter. Get out of the wind and wet as quickly as possible. Wind will remove any warmth that there is from the body and so it is important to protect from it immediately. Once in the shelter, get off the cold ground. Use anything as insulation but do not sit on snow or ice. Get a fire burning as quickly as possible to generate some heat. (Lighting fires will be covered in the next chapter.)

Protect children and babies especially from the cold as they will lose body heat much more quickly than adults. Look for natural shelters that will provide instant protection. You can then improve on them. Do not waste time and effort building shelters if nature has provided for you.

Be aware of the dangers of snow drifts and avalanches; from this point of view it is best not to shelter on cliff faces or under snow laden tree branches, unless they are supported. The weight of the snow could bring down the branches with you underneath them. If the branches are already down, however, and are resting on something, this will provide a good basis for a shelter.

Do not use the aircraft for a shelter in the extreme cold. Its metal fuselage will make it feel like a freezer inside and if there are any holes in the fuselage due to damage, it will act as a wind tunnel. This mistake has been made by survivors of aircraft crashes in Arctic conditions, many of whom have died before being found. The aircraft cabin as a shelter is not conducive to survival.

Try to get everyone into at least partial shelter immediately. People exposed to the wind will die very quickly. The cold numbs the mind; listlessness will set in fast so try to get your shelters built quickly, before you are either unable to do so or before you no longer care. The movement will help to warm you up and will ensure everyone is kept occupied, thus increasing morale.

Shelters

As we have said, the broken tree branch, supported on snow, is probably one of the best immediate shelters. The branches will provide a canopy and if this canopy is covered in snow, so much the better as this will form an insulating layer. These shelters can actually be quite warm because there is usually a build up of pine cones and needles that provide good insulation.

Cut away some of the underneath branches to give more room to get people under the shelter and try to break some of the branches at the edges to make them hang down. This will provide more protection at the sides. Do not break too many or you will dislodge snow. Try to dig a hollow out of the snow on the ground around the trunk and build it up on either side of the shelter to act as a barrier to the wind. Using pine branches or leaves as bedding helps increase insulation and comfort.

If there is no such broken branch available to you, and you feel it is safe to do so, try sawing a low branch, if you have anything that can be used as a tool, just enough that it falls down and make a shelter from this. Make sure that the fallen branch is supported safely at the bottom. The disadvantage here is that the snow will not have settled across the top of the branch but this form of shelter is better than nothing, especially as a temporary measure. To increase insulation, try tying the foliage of the tree together to make more of a canopy. A fire can be built under the tree, if there is room.

As an alternative tree shelter, use fallen trees. Lean logs and branches against a tree trunk that is lying on the ground to provide shelter. Again, scoop the snow away from the ground under the canopy to form a hollow and build this snow up on either side of the shelter to increase protection from the wind. Use branches as insulation from the cold ground and this will provide a useful shelter.

The second most common form of shelter is a snow hole. This is very efficient at providing a good level of protection but is rather labour intensive. If there are a large number of people to help work at the job, it is probably worth the effort but if there are just one or two of you, you may find it too much hard work. In this case, a simple snow trench would be better.

Snow Hole

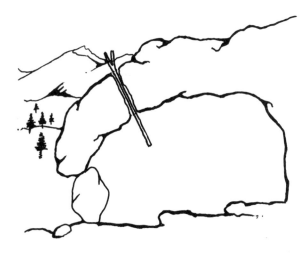

A snow hole.

The snow hole, or snow cave, requires very thick and compacted snow. There needs to be a depth of at least two metres and the best way to find this is often by digging into a drift. The principle is simply to dig into the snow. Dig a low entrance tunnel out first, on the lee-ward side of the shelter and then continue digging inwards and upwards. The key to making it stay there is to make sure the ceiling is domed.

Use the snow that is being excavated to build up the outer walls, or to make snow blocks that can be used to block the entrance once you are inside.

The sleeping area should be on a raised platform, above the level of the entrance, to protect the occupants from drafts and cold. On a raised level, you also take advantage of the fact that hot air rises and cold air sinks.

Once the occupants are inside, close the entrance with snow blocks. This is best done on the inside as it is less likely to freeze and trap you inside. Even if it does freeze, you will be able to move it far easier from the inside.

Ventilation is essential in snow holes so ensure that small gaps are left at the entrances. You need to let the carbon dioxide that has built up escape and let fresh air in. One idea is to make a hole in the roof, too, in order to ensure that enough air gets in. Do this by pushing a pole through the roof and moving it every so often to make sure it does not freeze up in the hole.

Light a fire inside the snow cave on one of the upper levels. As well as warming you up, this will melt the inner layer of the ceiling which will then glaze over as it cools, sealing and strengthening the roof. To prevent drips from making the floor wet, try to dig a slight channel around the inside of the cave, to take water away.

Snow caves are hard work to build and may be very difficult if the snow is packed down very hard. Tools will make the job much easier and may actually be essential in very hard snow.

Snow Trench

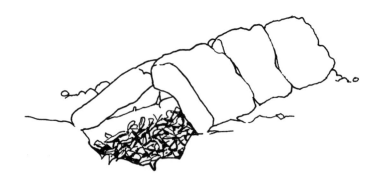

A snow trench.

This is a much simpler form of shelter but is smaller, suitable for only one or two people, depending on its size. It is simply a hole in the snow which can be enlarged and elongated to provide shelter. To offer any element of real protection, however, the trench must be covered and this is where the work comes in. Snow blocks placed on either side of the trench and lent against each other to form the apex of a roof are ideal but require more work. If the snow is too soft to do this, try using branches or some form of sheeting material. Aircraft parts may be a possibility here. Try to position the opening away from the wind.

Again, line the trench with vegetation or equipment to keep yourself off the cold, wet floor and you should be able to get a good night's rest in one of these shelters.

Igloo

An igloo is the top-of-the-range snow shelter but requires a lot of work, a lot of effort and good tools. This construction also takes time. Compacted snow is necessary to build an igloo. Simply cut the snow into large blocks and build them up in a spiral form, making sure each block leans slightly inwards. Build an entrance tunnel at ground level on the lee-ward side and, as in a snow hole, build a raised platform to sleep in.

This form of shelter will almost certainly involve too much effort for those who are in a forced survival situation. It also depends on being able to improvise the tools to cut the ice blocks.

When using snow shelters, there are certain considerations to bear in mind.

* Ventilation must always be considered.

* When entering snow shelters, make sure all snow is thoroughly shaken off clothing because you do not want to bring it inside the shelter. It will melt and soak into clothing as it warms up. The aim must be to stay as dry as possible at all times.

* Keep tools, if you have them, inside the shelter in case you have to dig your way out.

* Make sure someone tends to the fire, if there is one, at all times. It will be hard enough to light a fire in the first place. Do not lose it once you have got it going. (Lighting fires will be covered in the next chapter.)

In cold regions that are not quite as extreme as Arctic regions, consider building a shelter using branches and leaves. The diagram below shows how a fire has been built close to a shelter, with a barrier to prevent the heat from being lost. This sort of shelter requires some work but is worth it, especially in a prolonged survival situation, as it does afford some protection and warmth. (See the following chapter on fire building.)

**A shelter using branches and leaves. Note the guard
to prevent the heat from the fire escaping.**

Take the opportunity of being in the shelter to take off clothes and dry them. This is important as dry clothes are so much better at insulating. Also important is taking off your clothes occasionally, crumpling them up, rubbing them together and shaking them out before putting them on again. The compression of wearing clothes reduces the fluffiness of the fabric. These flattened fibres then lose their air traps and therefore their insulating quality. Do not stay unprotected in the shelter, however. This is the opportunity for you to dry or shake out clothes but you must still keep as well covered as possible as you will be sitting still and will feel cold quickly.

Wear shoes and boots in bed to prevent your feet, and your boots if they are wet inside, from freezing.

When in the shelters, preserve body heat as much as possible. As in a life raft, huddle together and keep toes and fingers moving to maintain circulation. When sleeping, again keep together. Sleep back to back to share warmth. The more people there are in a shelter, the easier it will be to keep warm. This is when it becomes imperative that ventilation is checked on a very regular basis.

Be aware, when working hard to build these shelters, that people do not overheat. Although it is important to trap air inside clothing and not let it escape when it is very cold, it is important to let some hot air out that has formed due to excessive exertion. If this is not done, perspiration will cause clothing to become wet. When the exertion ceases, this will then cool and will be dangerous. It can even freeze on the body. The other danger is dehydration and fluid loss. This may sound ridiculous in such a cold environment, but with so many layers of clothing on, and such rigorous exertion being carried out, perspiration can be extreme. This leads to a lack of body fluids. Remember, water is every bit as important in cold climates as it is in hot.

Looking after your body is another important way of protecting yourself in the extreme cold. Try to ensure that the body is kept as dry as possible. Armpits, crotch and feet are the most important areas to consider because skin that is wet for some time can crack and become easily infected.

Although cleaning your teeth may be the last thing on your mind, it is important to try and rinse your mouth out after eating or use a toothpick because gum infections are far more painful in the cold.

Breathe through the nose in cold weather. The nose warms air as it travels to the lungs whereas air entering the body through the mouth goes to the lungs cold. Those who breathe through their mouth are far more likely to succumb to chest infections and dry coughs. Smokers will suffer badly in the cold and are very susceptible to chest infection. So is anyone else in the vicinity of their smoke.

Lavatory Arrangements

Lavatory arrangements in Arctic regions are slightly different than in other land survival situations. The best advice is to relieve yourself inside the shelter. This is the only safe way of ensuring that too much body heat is not lost. Use containers and then remove all waste along with any other rubbish to a suitable distance from the shelters. Ensure that this area is well marked so you do not take snow and ice from near-by for water and ensure that you do not put waste near to running water that you may wish to use for drinking further down-stream.

Water

Water acquisition is very important in Arctic conditions. Although the temperature is very cold and you are surrounded by snow and ice, the atmosphere is in most places very dry and it is easy to overheat when wrapped up warm and undertaking exertion. It is vital, therefore, to drink plenty of water to retain body fluid. Force everyone to drink, even though they may not feel inclined to.

Water brought from the aircraft is obviously the easiest to use although this will freeze very quickly so you are likely to end up having to melt it over a fire before you can drink it anyway.

Additional water will have to be found. This is easy enough in these regions where snow and ice are abundant but will require a fire and some form of container for melting.

Do not use discoloured snow as this is not clean. Ice is better than snow for producing water as it is more dense. As the ice melts, it will provide the equivalent amount of water whereas snow is full of air. 17 cubic inches of snow will need to be melted to provide just one inch of water. Ice also requires less heat to melt it.

Put the ice or snow into a container over a fire. If the sun is shining, you may get away without the fire. If using the sun's rays, spread the ice or snow as thin as possible to increase the melt.

Melt only what you need and use all the water you have produced. If you let it re-freeze, you have wasted fire fuel and will have to light another fire to re-thaw it.

Snow and ice should be safe to drink without purifying, as long as it was white or clear before melting but if you can purify it, so much the better. As in every survival situation, illness could be the final stage to someone dying.

In slightly less severe conditions, look for a natural spring or stream as your water supply. You may have to cut through the ice to be able to access it but once you have done this, cover the hole up with boarding or snow blocks so it will not re-freeze. You then have a constant water supply. Water from a natural spring should be safe to drink. The only danger is a dead animal further upstream. If you have them, use purifying tablets or boil the water.

It is a good idea to have hot drinks, simply because they will warm the body up. In fact, icy cold drinks will cause stomach cramps and so should be avoided if at all possible. If very thirsty, snow can be eaten but should be compressed so that water drips are consumed rather than snow itself. Try only to do this in very small quantities because again, the cold will cause cramps and could even lead to hypothermia.

NEVER allow anyone to drink alcohol in this situation. It vasodilates, ie, widens the blood vessels and will simply act to flow the warm blood that has collected protectively around the vital organs away from them. In severe exposure cases, it can cause quick death.

Food

In such cold weather, food can be instrumental in keeping people alive as it provides essential energy which is nothing more than warmth. Hot drinks will warm people up but food will provide calories. The glucose tablets and survival bars carried in most survival kits will be ideal but only eat if there is a sufficient supply of water.

If you do not have access to survival rations and there is nothing on the aircraft that can be used, think about eating vegetation. Test it using the method outlined in the desert survival chapter. Meat is the other alternative but this involves a lot of work trapping, snaring and killing an animal, before preparing it for eating. Trapping and killing large animals presents a great degree of risk to yourselves while small animals will not satisfy many people.

15   Lighting Fires

Building and Lighting Fires

This skill is useful for a variety of reasons. A fire can be used for warmth at night, to keep insects at bay and to boil water. It is also useful as a central focus - something that everyone can sit around that will increase the sense of security and so improve morale.

The desert is probably the easiest place in which to start a fire as it has the driest atmosphere although there may be the least to burn. The task of lighting a fire is considerably harder than most people ever imagine.

Remember the basic rules of fire. There are three components of a fire and without any one of these, fire will not happen. These are HEAT, FUEL and OXYGEN. This means that you need something to burn, enough heat to make it burn and adequate ventilation.

Prepare your fire. Choose a location for it. If you want this to be in the middle of the 'camp' then choose a safe and level spot away from vegetation and any equipment or shelters. You do not want to lose valuable equipment by burning it!

Make sure you have enough fuel. This should be divided up into three categories: tinder, kindling and fuel. The tinder and kindling are used to burn initially in order to get the fire going. Once burning successfully, the fuel will then catch alight. Be aware that nothing is burned that may give off toxic gases. A shortage of large items to burn such as logs may be a problem in the desert; be careful what is used a substitute.

Good tinder should need only a spark to set it alight, especially in dry conditions. Tinder should be easy to find in the desert. Dry grass, wood shavings, bird down, cotton wool and charred cotton or linen are excellent. If anyone in the party has a cigarette lighter then it should be easy to provide charred cotton or linen. Tinder is a good item to keep in survival kits as it must be completely dry.

Use your initiative, though, as there are many other items that can be used. Matted hair, certain photographic film, rope, clothing and bandages will all burn and one of the best sources of tinder is the cotton wool from a tampon or sanitary towel. This is so effective that a tampon is now standard issue in the survival kit for many armed forces around the world.

In addition, certain liquid substances can be used to increase the combustibility of the fuel. These include lip salve, oil, insect repellent and aviation fuel. Aviation fuel, although it evaporates quickly in the heat, will be ideal to enhance the fire.

Vegetation and earth around the crash site will still be soaked with enough fuel to be a useful combustion aid.

The slightly larger twigs and sticks will form the kindling. As the fire burns, add larger items for burning.

Building the Fire

Make sure that enough material is collected before attempting to light the fire. Fires usually take more material than you think. Do not stack too much material up too quickly, though, as you will suffocate the fire and it will die.

Lighting the fire on a base of logs can help in wet or marshy conditions.

If it is windy, think about protecting the fire by lining the area in which you are going to light it with rocks or aircraft parts. This can be a good idea anyway, once the fire is established because a fanned fire burns far more fuel. If you can control the ventilation, you can conserve burning material. If there is nothing else, use your body as a wind shield.

The other benefit if having rocks around a fire is that once the fire is out, the rocks retain some heat and act as radiators. An example of how this has helped save lives occurred in the Cessna crash in the Nevada mountains in 1976. The two survivors used raw aircraft fuel to burn on a pile of rocks. Once the fuel was exhausted, the rocks were placed in the shelter and used for warmth.

How to Start a Fire

For this you need a heat source. This shouldn't be a problem in the desert. Again, be inventive. Spectacles held in the sun can be used to ignite tinder. Use a cigarette lighter or a match if someone has one. Waterproof matches are available and should be kept in the survival kit. A candle should also be part of the kit. This has many uses. Once it is alight it can provide light, start fires and can be used over and over again. In addition, the wax from it can be stored and either made into a second candle or used to line and waterproof containers.

A candle can be an invaluable tool for fire lighting. Lighting a fire in the way shown above can encourage the combustion of damp wood.

Direct flame will light tinder immediately. If you cannot use direct flame and have to resort to heat, for example, the sun's rays through glass, you may achieve only a smouldering combustion that does not look as if it is going to burst into flames. By gently blowing and fanning the smouldering part, however, you will increase the ventilation and oxygen to the area until ignition takes place.

Do not be afraid to resort to the old method of flint and striker if you do not have easier means with which to start a fire. This is hard work but can be very effective if you persevere. Fire kits that should be part of the survival kit will contain a flint and a steel striker that has hardened teeth. Use this to generate a spark that will ignite the tinder, for example, cotton wool.

The Jungle is the hardest region in which to start fires because of the amount of moisture. Careful searching, however, will locate some dry tinder in these regions. Look for dead, dry branches and leaves which can be used to start a fire.

16 First Aid on Land

This chapter does not intend to teach general first aid. Instead it aims to cover the types of illness and injury typical of air incidents and, particularly, of the different survival situations. It also takes into account that medical help is not just a 'phone call away and that the survivors will have to cope with nothing but the resources and knowledge that they have between them. Medical personnel may be on the flight but this can not be counted on. Even if they are, they may not have survived and are certainly unlikely to have any equipment with them.

The first rule in wilderness first aid is 'be ruthless'. Select those who you help and do not waste time on those that cannot be helped.

There are six categories of assessment in wilderness first aid. These are:

* Dead.
* About to die.
* Seriously injured and unstable.
* Seriously injured but stable.
* Injured badly but will survive.
* Injured slightly and able to help themselves.

Assessing who to help is the difficult part. There will may other tasks to be done simultaneously, if not prior to, helping the injured. The group must come before the individual and it is more important to get the majority out of the freezing wind in the Arctic or out of the relentless sun in the desert than it is to see if you can revive a few injured passengers who may not live anyway. Do not ignore the injured but do not let others die because basic survival priorities are not adhered to.

Use common sense. See to the injured as soon as possible or split tasks if there are enough of you. Once those who have got out of the aircraft have been taken to safety, and this will probably involve getting them into shelters, go through the wreckage and see if there are any injured survivors that can be helped.

This can be a daunting task. However horrific the injuries, do not assume immediately that they are dead. Injuries can be deceptive. What looks horrendous and very bloody may not be life-threatening. On the other hand, someone may be very dead with no outward signs at all.

The first rule of first aid in any situation is 'keep calm'. The second is 'assess the danger to yourself and others before acting'. See if there are any medically trained people in the party who are well enough to help.

Feel for breath on the back of your hand or on the cheek. One way of assessing whether there is any neurological activity is to put the hand over their mouth quite firmly. A reaction to this, however small, shows life.

The first action with any injured survivor is to check for the airway and a pulse. If there is neither, you have to decide whether to try and resuscitate. If there are several passengers in this condition, you cannot spend time on them all. If there is just one, it may be worth a try as long as other essential priorities of survival are not neglected. Think about the time that they have been like this, too. If it is more than a few minutes, the chances of reviving them, with no medical help within hours, are very slim. Do not waste your energy on things that will not help the overall survival effort.

If the person is alive, keep the airway clear at all times and monitor them very closely. Move them out of danger. This is vital so do not be afraid of injuring them further by moving them. Keeping them in danger may simply kill them. Danger includes being in the aircraft in the heat of the desert or the cold of the jungle, being exposed to the freezing cold, the searing heat or a plague of poisonous insects.

Get them to shelter fast and keep them warm in cold climates, cool in hot climates. Put them on the priority list for fluid but do not waste vital water supplies on them if they are very weak and their chances of survival are slim.

There are two ways of trying to asses what is wrong. One is by asking the patient what they feel; where they are hurt etc. Whatever facts the patient can tell you are called 'symptoms'. Anything you can see without the patient telling you, ie swelling, face colour, sweating, are called 'signs'.

Shock

Shock is one of the most fundamental conditions that you should be aware of. It can happen after many different types of injuries as it is caused by a lack of blood in the body. This lack of blood means a lack of oxygen getting to the brain, heart and lungs which can seriously impair physical and mental functions. The symptoms include cold, pale and clammy skin, rapid breathing, a rapid but weak pulse, sweating and often thirst. Treatment includes preventing the cause, by stopping the bleeding; making the patient comfortable, keeping them calm, being reassuring, loosening clothing, keeping them warm and as always, monitoring consciousness levels.

Do not take this condition lightly; it can easily result in death.

Bleeding

There are two types of bleeding: arterial, from the artery and venous, from the vein. They are quite easy to differentiate. With arterial bleeding, the blood will be a much brighter red and will be pumping under pressure. This is because the arteries carry blood away from the heart and so have the pressure of the heart's beat behind them. The blood is bright red because it contains all the oxygen picked up from the lungs. Typically, it will be spurting out.

Venous blood, on the other hand, is blood that is travelling back to the heart. It has little oxygen in it and so is a much darker red. It is not under as much pressure and so will not spurt.

Serious bleeding of both types should be treated by applying pressure and elevating the offending area.

Apply pressure and elevate.

Apply dressings firmly and apply more over these if the blood continues to soak through.

If the bleed is arterial and both direct pressure and elevation do not work, try pressure points. This is easiest to do on a limb. Pressure points are found where an artery runs across bone close to the skin. There is a pressure point on each limb. On the arms, they are on the brachial arteries, running down the centre of the inside of the upper arm. On the legs, they are on the femoral arteries, which run down the inside of the thigh.

In a survival situation, away from medical help, the tourniquet is always there - but only as a last resort and only ever for use on limbs. This method of stemming bleeding is no longer recognised by medics. If you can not stop bleeding any other way, take a strap of any kind such as a belt, tie or strip of cloth and wrap it tightly, twice around the limb above the wound. Tie two overhand knots. It is vital that this tourniquet is loosened every 20 minutes, left loose for a full minute and them re-applied. Be aware that the use of a tourniquet may well result in the loss of the limb.

Be aware, too, of the dangers of toxins developing in the limb that has had the oxygen supply cut off and of these toxins flooding back into the body, causing severe problems and even death. This is why it is essential to release the tourniquet on a regular basis.

Avoid direct pressure on a wound if there are foreign bodies in it - instead make a ring around the wound and cover it because infection can get in very quickly. Never remove foreign bodies from wounds, especially in the trunk of the body.

Watch out for internal bleeding - it is easy to go unnoticed and can be silently deadly. The principle symptom is shock which increases in severity as time passes. Treat for shock and monitor at all times; unconsciousness can occur very suddenly.

Always be aware of the dangers of blood borne pathogens and take the necessary precautions, for example, when applying pressure to a wound, ask the casualty to hold the wound themselves. Wear gloves, if available and try to avoid contact with body fluids.

Fractures

Diagnosing fractures is not always easy. Look for pain, swelling and any abnormal shape of the area. There are generally two types: a simple fracture in which the bone is broken inside the skin and a compound in which the skin has been broken by the bone. In a severe case, the bone can be sticking out.

Treat these as you would in any other situation. Immobilise the fracture as much as possible. If you are not sure whether broken or not, immobilise it all the same. Improvise a splint if there isn't one in the first aid kit or, in the case of a leg, use the other limb.

The SAM splint, manufactured by The Seaberg Company, Inc, is a great all purpose splint to carry in survival kits. It is lightweight and malleable and can be used over and over again. It is made of rectangular strips of closed cell foam material and aluminium which is less than 20 thousands of an inch in thickness. These splints are flimsy but can be made very strong by putting rigid structural bends in them.

A SAM splint.

Triangular bandages should also be carried on first aid kits as these have numerous uses.

In compound fractures, where the skin has been broken open, watch for infection.

Skull fractures are often characterised by bleeding from the ears, nose and mouth. Cover the ears with a clean cloth to prevent infection and, presuming the patient is breathing, lie them in the recovery position. Do not let them get up and walk around, however well they feel. Keep the patient warm.

If a neck injury has occurred, keep the patient as still as possible and immobilise the neck very carefully. The SAM Splint can also be used for this job. Take care that the patient's breathing is not impaired when securing the splint.

Be aware of fractures in the upper legs. Apply traction to such injuries to try to prevent the bones causing internal damage. If bones break, especially the femurs, the muscles are so strong that they will pull the bones out of place very quickly. A broken femur can contract and can sever the femoral artery - all internally if the skin is not broken. A patient will lose four pints of blood into such an injury. If they have broken both femurs, they can lose eight pints of blood and this will kill them very quickly.

If you suspect a chest injury keep the casualty in a semi-sitting position. This puts the least strain on the heart and keeps the fluid at the bottom of the lungs, allowing the greatest capacity for gaseous exchange. If laid down flat, this fluid would seep throughout the lungs.

A flail chest occurs when something has punctured a lung. A broken rib could easily do this. The signs of this condition are pain, breathlessness or noisy breathing and the patient may be coughing up blood. Lean them slightly to the side where the injury has occurred to prevent fluid from the injured lung draining into the uninjured one and keep them in a semi-sitting position.

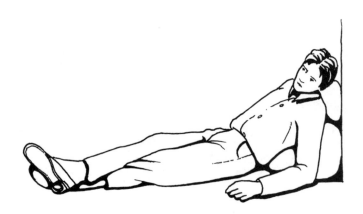

If you suspect a chest injury, keep the casualty in a semi-sitting position.

Dislocations

These are extremely painful and debilitating. If there is a medically trained person in the party, get them to try and put the dislocation back in. This will provide instant pain relief and it will then be easier to immobilise the limb. The longer dislocations are left out, the harder it takes to put them back in. The general rule is that it will take the same amount of time to put a dislocation back in as it has been out. If dislocations are left out for longer than two hours, they can cause damage. As with the femurs, the muscles of the body will try to put the bone back and will pull it into the wrong place. This can trap blood vessels to dangerous limits.

Burns

There are generally two types of burns. Full thickness burns mean the burn has penetrated deep and will have damaged or destroyed tissue and nerve endings. The skin will look white or charred and there will be little pain at the burn site because of the nerve damage. There may be severe pain, however, where the damaged tissue joins less damaged tissue.

Partial thickness burns are very painful, show red skin and are often covered in blisters.

Immerse burns in cold water to take the heat out of them. This will reduce the swelling and prevent further burning from taking place. Try not to touch the burn with the fingers - the area will infect very easily. Do not burst blisters for this reason and do not pull clothing off from over the burn - it will take the skin off with it.

Cover the burn with a clean dressing - removing the burn from the air usually provides immediate pain relief - and put this person on the priority list for water to reduce the risk of dehydration. The kidneys will be at risk if fluid is not replaced. Give water in small quantities by mouth.

In the desert, the sun can also cause serious burns. Use sun cream, if available, and keep the whole body covered. The sun can burn far quicker than you think, even through thick cloud.

Conditions to be Aware of in Cold Climates

There are two conditions that survivors in cold regions need to be aware of. These are hypothermia and frostbite. (See Chapter 9, First Aid at Sea, for further details on hypothermia.)

Hypothermia

Hypothermia is a condition whereby the body loses heat more quickly than that heat can be replaced. A person is said to have hypothermia when the core temperature of the body drops from the norm of 37 degrees Centigrade (98.4 degrees Fahrenheit), to 35 degrees Centigrade (95 degrees Fahrenheit) or below. It does not have to be that cold for prolonged exposure to the cold to result in hypothermia.

Hypothermia must first be recognised, in time to be treated, and you can guarantee that the person suffering from it will be the last to be aware of it. In fact, in the later stages they will be feeling very sleepy, not particularly cold and will just want to nod peacefully off to sleep. If they are not spotted before they do this, they will not wake up again!

The signs are fatigue, slurring of speech, poor judgement, uncontrollable shivering, pale skin and being very cold to the touch, impairment of memory, loss of awareness and a slow pulse. Once the temperature drops below about 32 degrees Centigrade (90 degrees Fahrenheit), the condition is extreme and the shivering mechanism fails. At this point, the casualty is close to dying unless action is taken immediately.

Watch out for the quiet ones. If they are cold, quiet and tired, be aware. They may have hypothermia. It can creep up on a member of the survival group without you being aware, especially if you are busy with may other tasks.

It is essential to warm the person by any means - body heat is one of the best insulations so huddle close and hold them. Remember to cover the head as it is one of the greatest areas of heat loss.

Warm the casualty from the core outwards. Keep the casualty still. The key is to warm them slowly so that the little warm blood that they have, which has remained in the central organs, does not flood away from these organs, killing the person. Do not, then, rub the limbs or put them too close to a fire. Never administer alcohol - is will cause the blood vessels to dilate and will encourage that warm blood to flood round the body faster, leaving the vital core organs in order to do so.

Wrap them in any spare clothing or survival blankets and monitor them. Put them between other people in the shelter for warmth.

Encourage blood flow to the heart, lungs and brain by lying the casualty down in a slightly head down posture - in the recovery position but with the legs slightly raised.

As in a sea survival situation, people can suddenly go hypothermic after some time. Do not assume that because they were not a serious case at the beginning that they will not suddenly become so.

Frostbite

Frostbite is the freezing of some parts of the body. It occurs at the extremities of the body, such as the fingers, toes and ears and is caused by exposure to subfreezing temperatures, especially freezing wind. Affected areas appear white; they will be numb and feel hard to the touch. Preventative methods include wearing protective clothing and keeping the fingers and toes moving to maintain that all so important circulation. Also wrap the ears and cheeks in clothing as these, too, can be susceptible to frostbite. Making faces to keep the mouth and cheek muscles moving may also help stem the onset of this condition.

Put affected toes and fingers into a warm place, for example in the groin or under armpits. Do not squeeze or rub. If the survival situation is likely to be prolonged, it is inadvisable to thaw these parts of the body if they are only going to re-freeze.

Do not underestimate frostbite. Many a finger or toe has been lost through this complaint and the exposure to the cold need not be prolonged for this to happen.

For example, if the wind chill factor is -45 degrees Centigrade, (an air temperature of -10 degrees combined with a wind speed of 15 mph) exposed flesh will freeze in only 60 seconds!

Conditions to be Aware of in Hot Climates

Heat or sun stroke, heat exhaustion, sun burn and dehydration are all conditions to be aware of in hot climates. Be aware, though, that sun burn can occur in any climate where the sun is strong and dehydration will occur whenever a person is not consuming sufficient water.

Heat Exhaustion and Heat Stroke

Both these conditions can occur in hot climates and can be very serious. They are caused by too much exposure to high temperatures and are exacerbated by high humidity. The young, elderly, ill or intoxicated will be most at risk.

Symptoms of heat exhaustion are sweating, dizziness, nausea, rapid breathing and brief loss of consciousness. This must be treated straight away by cooling the person down. Splash water over them, cover them in damp clothing, keep in the shade and monitor them. Give water to drink if available.

The next step manifests itself in heat stroke. By this time, the person has stopped sweating and is in serious danger of sustaining irreversible brain damage or even death. The body cannot naturally lower its core temperature any more because the sweating mechanism has broken down. They must be cooled very quickly by sponging with cool water. Get their clothing off to do this so the water is next to the skin. Be aware of exposing vulnerable skin to the sun for too long but act quickly because this is critical.

Dehydration

The human body is made up of 75 per cent water. It helps keep the body at a constant temperature and keeps the kidneys functioning. A 25 per cent loss of body fluid can be fatal.

When seriously dehydrated, the pulse becomes faster because the heart has to pump harder to get the thicker blood round the body. Skin, when it is pinched, does not spring back when released.

Water consumption is the only remedy for this condition. Getting out of the heat and sun in hot climates, however, can also help.

Disposal of the Dead

Macabre as it sounds, there must be a plan for disposal of the dead. Leaving bodies unburied generates both a hygiene and morale problem. Bodies should be taken well away from the survival camp and buried where they will not pollute water sources. Burial sites should be clearly marked for later recovery and any personal items, such as watches, rings or pocket books, should be secured and kept safely.

Although considered morbid by many, the body should be stripped of clothes which may be useful to other survivors.

17 Signalling and Rescue on Land

In any survival situation, rescue must be your ultimate aim. In fact, short of keeping people alive long enough to be rescued, it is your only aim.

Throughout the survival ordeal, whatever challenges you are faced with such as finding shelter, building fires and dealing with the injured, you must always be aware of potential rescue attempts. Never assume that rescuers will come to you. As in a survival at sea situation, rescuers may be aware of your position; that does not mean that they will be able to find you immediately. Keep in mind all the incidents that have been outlined in this book and difficulties rescuers have had in finding survivors. You must help them all you can.

In many ways, a land survival situation is far easier than a sea survival one. For a start, you can stay within the vicinity of the aircraft and once you have found shelter, there is no need for you to move away from this spot. This makes the rescuers' job much easier. They will begin to search the area around which contact was lost with the aircraft. If a rough crash landing position was given, this gives the rescuers a pretty good start point.

On land, the aircraft will still be there. It may have burned out but signs of the crash will still exist. In the sea, it could have, and probably would have, sunk without trace. As survivors in a raft, too, you would probably be nowhere near the landing spot after a few hours.

So the rescue job is far easier on land. This is assuming, of course, that the crash landing was planned and that the flight deck have been able to give some notice. But do not allow anyone to be complacent. The Air Inter Airbus A320 that crashed in France did so while on its final approach leg to land at Entzheim Airport, Strasbourg. It was no more than ten miles from the runway when it went down. Despite this, the rescuers took several hours to find the wreckage; not as long as in other incidents but it is thought about six people died of exposure in this time.

The Varig Brazilian Airlines crash is another example. Although survivors had a working ELT, and rescuers knew where to start looking, it was 44 hours before the crash scene could actually be located.

In another incident in 1971, again in the forests of South America, a Lockheed Electra crashed after suffering a catastrophic structural failure. The wreckage was scattered over 10 miles in the mountainous, forested terrain in Peru. The wreckage was not found for two weeks. A survivor, a 17 year old girl, was found ten days later by hunters in the jungle. She had been walking through the forests for this time. When the wreckage was eventually found, there were no other survivors.

Evidence shows, however, that at least 12 others did survive the crash and so died while waiting for rescue.

In yet another incident in the South American jungle, in Brazil, a Cessna carrying five people made a forced landing after suffering engine problems. The pilot of the aircraft was in communication with another aircraft shortly before making the landing. The pilot of the Cessna turned on his Emergency Locator Transmitter (ELT) before the accident and so the search aircraft was able to use this as a homing device. It still took 25 hours, though, before the crashed aircraft was found through the thick vegetation. There was one survivor.

These incidents should outline the importance of maintaining a constant watch for signs of rescue and being ready to make yourself as visible as possible very quickly. Do not assume you will be found immediately.

Many of the techniques and certainly the equipment that you have to use will be similar to that in a sea survival situation. On land, however, you will also have the advantage of many natural items that you can use to make yourself conspicuous.

It is vitally important to give as much information, presuming the landing is planned, while still in flight. Switch the aircraft's transponder to the emergency frequency of 7700. Turn the ELT on as soon as possible. This will then start to transmit. Once the crash landing has taken place, one of the priorities is to take all the signalling equipment off the aircraft.

The majority of this equipment was covered in Chapter 10, Signalling and Rescue at Sea. Refer back to this chapter for further details, especially on the COSPAS-SARSAT system. This chapter aims to outline the use of signalling equipment on land and other ways of signalling, using natural material.

Emergency Locator Transmitter (ELT)

This is one of the most valuable signalling devices. If the aircraft survival packs contain hand held ELTs or Personal Locator Beacons (PLBs), make sure these are brought from the aircraft too. There have been many incidents in which the ELT in the aircraft tail has failed in the impact, thus rendering it useless. The more beacons you have, the more chance there is of one of them working. Because they have a limited battery life, too, more than one beacon will allow transmittal for much longer.

Some beacons require submersion in water to activate them. On land you do not have the advantage of being surrounded by sea and so have to find water. If there is a stream, lake or shore close by, put the beacon in this but ensure that it is firmly tied to something or you could lose it. If there is not such a water source, do not use valuable drinking water. Dirty water or even urine will do the trick. Do make the beacon a priority, however, even if water is very scarce because it may be the one thing that alerts the authorities to your predicament.

Cellular Telephones

These are worth using in any survival situation, if any one has one and if you can get a strong enough signal on it. It is possible for the authorities to triangulate your position with a cellular 'phone by using different masts.

Signalling Mirror

As in the water, this is an invaluable piece of equipment with a very good range of visibility. It can be used anywhere as long as the sun is out. Desert or sunny Arctic conditions are ideal. The jungle is not so good unless you can find a clearing. Even then, you will have only the limited time that it takes the aircraft to fly over the clearing to use it effectively.

The signalling mirror is good, especially if someone is aware of your predicament and is out searching for you. On land it may be slightly less effective in that a flash from a mirror can more easily be confused as a flash from a window or a car and ignored. At sea, where a pilot flying over it knows there is no land for miles, a flash like this will certainly attract attention. It is still worth its weight in gold, however, as it takes no effort to use.

For details on how to use a signalling mirror, see Chapter 10.

Sea Dye Marker

This is only useful on land if you are stranded close to a beach or water source such as a lake. If this is the situation, it may be particularly useful in a jungle situation, where rescuers are not going to be able to see through the tree canopy. Put the marker into the sea along the shore line and this will at least give rescuers an idea that the survivors are close by. In the jungle, it is worth travelling a little away from the crash site in order to do this. If the group is large, let a small group make the journey to do this. They can then alert the services to where the larger group is.

Flares

These should be used in the same way that they are in a raft. Only fire them when you are sure that someone is looking for you. They are in limited supply and so you do not want to waste them. See Chapter 10 for further details on flares.

Whistle

This item of equipment could be useful to let rescuers know where you are but only if they are searching for you on foot. Again, the jungle is probably the most relevant place to use such a piece of equipment.

If one member, or a small group, of the survival team leaves the main party for any reason, a whistle is a useful thing for them to take with them, in case they become lost.

Whistles could also be useful in 'rounding up' survivors who may be scattered over a wide area. Remove whistles from the lifejackets. It certainly cannot hurt for each member to have one with them.

Chemical Light Sticks

These are useful at night to pinpoint your exact position. Only use when rescuers are in the vicinity and are looking for you.

Flashlight

Always a useful device as it can be used for visibility at night and for signalling in the dark. Spare batteries should always be in the survival pack.

Strobe

Most life rafts are fitted with strobe lights but it is also possible to get hand held strobes. These flash very brightly, especially at night and have an eight or nine hour life. Take advantage of the raft strobes if you carry no others. You may be using the rafts for sheltering so try to position them so the strobe can be seen from the air. If using hand held strobes, conserve the batteries until help is imminent.

Fire

Fire is an invaluable signalling device in any land survival situation. The smoke is useful during the day and the flames at night. To this end, try burning a smoky fire by using slightly damper wood during the day. See Chapter 15.

Make yourself Conspicuous

Make yourself as conspicuous as possible. Improvise - there is so much around you that you can use. One rule to remember is that nothing in nature is in perfect circles or straight lines. Circles and straight lines, therefore, show up very well from the air. If they are big enough, they have even been known to be seen from a satellite.

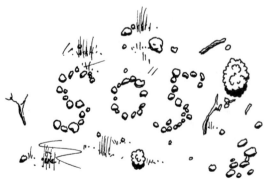

Use whatever is around you to attract attention.

Use whatever you have available to you. In the desert, for example, you can use rocks to mark the ground. Make big circles or straight lines or even make an SOS shape on the ground. Things like this show up very well from overflying aircraft.

Use contrasts. In snow, for example, lay anything down that is bright in colour. The orange SEE/RESCUE streamer that is mentioned in the Sea Survival chapters will be ideal to show up against a snow background and will be spotted easily from the air. A fire will do the same.

**A SEE/RESCUE streamer shows up like a beacon.
Here is one being used in an Arctic region.**

The jungle is undoubtedly the most difficult region in which to signal effectively. Try to find a clearing from which to use signalling equipment or you really will be invisible from the air. One invention, used a lot by the military, is a bright orange balloon on a very long string that can be raised up through the dark green tree canopy. This sticks out like a sore thumb and can be seen from quite a distance.

18   Conclusion

Survival is not easy but it is possible. As can be seen, there are many elements that add together to provide success in survival situations and almost all of these involve the crew.

Certainly, as far as aircrew are concerned, survival is not an area that can be left to chance. Prior knowledge, and, if at all possible, experience of practical survival situations within a training environment, are the very least that every crew member should be equipped with to be able to cope should such an incident happen to them.

Equipment is the other factor that makes the difference between life and death and yet is so often seen as little more than a money-wasting commodity. It is so common, for example, for General Aviation pilots to depart on a flight over water and not take any sort of life-raft with them. Airlines are not required to carry liferafts; sliderafts are seen as adequate and in the US, detachable seat cushions are seen as an acceptable alternative to lifejackets for flights operating a 'close' distance from shore!

As another example, airline survival packs rarely contain enough water to supply half the number on-board with anything like even a mouthful and hardly anyone carries water making equipment.

It is no coincidence that those airlines who have experienced such an incident are the ones that are now acquiring equipment and pursuing survival training for their crew as if their lives depended on it. Bitter experience has taught them that the vast majority of victims of survival situations did not know what to do. They were not equipped, either in terms of knowledge, training or equipment or in terms of mental preparation, to cope and in many tragic cases, this lack of preparation has lead to unnecessary fatalities.

It is only when such an incident occurs so close to home that an appreciation that such things do happen is borne. Even then, unfortunately, scepticism among those who were not so closely involved is still rife.

Whether from the airlines or in General Aviation, aircrew have a responsibility to themselves and to their passengers to ensure that they are prepared to deal with any situation that arises. A pilot would not embark on a flight without enough fuel to last the trip; just because the chances of ending up in a survival situation are slim, it does not mean that it is acceptable not to bother to prepare for this eventuality.

Sources for Equipment and Training

The following represent a selection of companies offering survival equipment and training.

LIFERAFTS AND LIFEJACKETS

BF Goodrich
3414 South 5th Street
Phoenix
Arizona 85040-1169
USA
Tel: +1 602 232 4000
Fax: +1 602 243 2300

Beaufort Air-Sea Equipment Limited
Beaufort Road
Birkenhead
Merseyside
L41 1HQ
UK
Tel: +44 151 652 9151
Fax: +44 151 670 0958

Eastern Aero Marine
3850 N W 25th Street
Miami
Florida 33142
USA
Tel: +1 305 871 4050
Fax: +1 305 871 7873

Life Support Equipment Pte Ltd
Building 140
East Camp
Seletar Airport
Singapore 2879
Tel: +65 4820889
Fax: +65 4820962

RFD Limited
Kingsway
Dunmurry
Belfast
BT17 9AF
Northern Ireland, UK
Tel: +44 1232 606113
Fax: +44 1232 621765

Winslow Liferaft Company
11700 SW Winslow Drive
Lake Suzy
Florida 34266
USA
Tel: + 1 941 613 6666
Fax: +1 941 613 6677

EMERGENCY LOCATION EQUIPMENT

HR Smith
Street Court
Kingsland
Leominster
Herefordshire
HR6 9QA
UK
Tel: +44 1568 708744
Fax: +44 1568 708713

SEE/RESCUE
219 Koko Isle Circle
Suite 602
Honolulu
HI 96825
USA
Tel: +1 808 395 1688
Fax: +1 808 395 4470

Pains Wessex Limited
High Post
Chemins
Salisbury
Wiltshire
SP4 6AS
UK
Tel: +44 1722 411611
Fax: +44 1722 412121

SERPE-IESM
Zone Industrielle des Cinq
56520
Guidel
France
Tel: +33 2 97 02 49 49
Fax: +33 2 97 65 00 20

Search & Rescue
516 Eastpoint Tower
180 Ocean Street
Edgecliff
New South Wales
2027
Australia
Tel: +61 2 9363 9736
Fax: +61 2 9326 2713

SURVIVAL TRAINING

ANDARK
256 Bridge Road
Lower Swanwick
Southampton
SO31 7FL
UK
Tel: +44 1489 581755
Fax: +44 1489 575223

STARK Survival Co
6227 East Highway 98
Panama City
Florida 32404
USA
Tel: +1 904 871 4730
Fax: +1 904 871 0668

Fleetwood Offshore Survival Centre
Broadwater
Fleetwood
Lancashire
FY7 8JZ
UK
Tel: +44 1253 779123
Fax: +44 1253 773014

Warsash Maritime Centre
Southampton Institute
East Park Terrace
Southampton
SO14 0YN
Tel: +44 1489 576161

Helicopter Survival Rescue Services
81 Ilsley Avenue
Unit 7
Dartmouth
Nova Scotia
B3B 1L5
Canada
Tel: +1 902 468 5638
Fax: +1 902 468 3083

The Write Partnership
1 Bendles Cottage
Upper Clatford
Andover
Hampshire
SP11 7QE
Tel: +44 1264 356973
Fax: +44 1264 356974

LTR Training Systems
230 E Potter Drive
Unit One
Anchorage
Alaska 99518
USA
Tel: +1 907 563 4463
Fax: +1 907 563 9185

This is by no means a comprehensive list but gives a random selection of companies providing these services.

Bibliography

Aimi, C.A., Kneipp, U.T.D., Moreira, R and Fernandez, R.D.L. (1996), *Jungle, We Can Survive!*, Varig Brazilian Airlines, paper given at Southern California Safety Institute 13th Annual International Aircraft Cabin Safety Symposium.

Barton, Bob. (1997), *Outward Bound Survival Handbook*, Ward Lock, London.

Butler, Bill and Simonne. (1991), *Our Last Chance*, Exmart Press, Miami.

Callahan, Steven. (1987), *Adrift - Seventy-six days lost at sea*, Ballantine Books, New York.

Davies, Barry. (1996), *The SAS Escape, Evasion and Survival Manual*, Bloomsbury Publishing, London.

Davies, Barry and Beynon, Phil. (1987). *Survival is a Dying Art*, BCB International Ltd, Cardiff.

Dawson, Ron. (1995), *First Aid Procedures & Lifesaving Techniques and Outback Skills & Survival Techniques*, Microtek Publications, Boise.

Elder, Lauren and Streshinsky, Shirley. (1978), *And I Alone Survived*, Pan Books Ltd, London.

Frost, M.G. (1956), *Management*, The English Universities Press Ltd, London.

Gero, David. (1996), *Aviation Disasters*, Patrick Stephens Limited, Yeovil.

Horner, B.K. and Birchell, R.B. (1995), *Course Manual, Primary Aviation Survival School*, LTR Training Systems, Anchorage.

Job, Macarthur. (1994), *Air Disasters, Volume 1*, Aerospace Publications Pty, Ltd, Fyshwick.

Job, Macarthur. (1994), *Air Disasters, Volume 2*, Aerospace Publications Pty, Ltd, Fyshwick.

McManners, Hugh. (1994), *The Commando Survival Manual*, Dorling Kindersley, London.

Marsden, A.K., Moffat, Sir Cameron and Scott, Roy. (1994). *First Aid Manual*, Dorling Kindersley, London.

Penoff, R.E. (1985), *Aircrew Survival*, Department of the Air Force, Headquarters US Air Force, Washington.

Platten, David. (1986), *The Outdoor Survival Handbook*, David & Charles, Newton Abbot.

RMAS. (1978), *Serve To Lead* (An Anthology), The Royal Military Academy. Sandhurst.

Rolla, Colonel Carlos Alberto Pires. (1998), *Jungle Rescue*, Brazilian Air Force, paper given at Shephard Press, SAR 98 conference.

Robertson, Dougal. (1973), *Survive the Savage Sea*, Elek Books Limited, London.

Stewart, Stanley. (1994), *Air Disasters*, Barnes & Noble Books, New York.

Watney, John. (1987), *The Royal Marines Commandos Fitness & Survival Skills*, David & Charles, Newton Abbot.

Wiseman, John. (1996), *The SAS Survival Handbook*, HarpersCollins Publishers, London.

Index

Air China 11, 12
Air Florida 11, 12, 97
Air Inter 71, 119
Alcohol 4, 53

Bailer Bucket 45
Boarding a Raft 31, 43
Brace Position 9, 15
Briefing 14
Buoyancy Aid 17

Chemical Light Sticks 47, 64, **122**
Children 22, 25, 100
Clothing 15, 28, 35, 53, 55, 71, 74, **77, 78,** 91, 92, 98
Controlled Flight into Terrain (CFIT) **71**
COSPAS/SARSAT 59, 60, 120
Crew Resource Management (CRM) **14**

Dental Floss 47, 85
Distilling 79, 80, 93
Ditching 8, 10, 12, 13, 14, 17, 21, **28, 49**
Drogue 33, 43, 67
Dunker 17, 19, 20
Dutch Antillean Airlines 8, 10, **14, 17**

Ethiopian Airlines 10, 11, 14, 17
Evacuation 2, 5, 12, 13, 15, 16, 17, **18, 21,** 31, 71, 73, 74, 91, 92, 99
Eyesight 78, 99

Filtering Water 83
Fires 70, 72, 86, 100 - 103, 107 - 109, 119, 122
First Aid
  Administration 27, 36, 70, 84, 110 - **118**
  Bleeding 55, 111, 112, 113
  Burns 49, 54, 115
  Cardio Pulmonary Resuscitation **27, 50, 51**
  Cramp 54
  Dehydration 37, 55, 73, 83, 93, 117
  Drowning 17, 19, 21, 22, 23, 49,

  Fractures 27, 49, 113, 114
  Frostbite 55, 116, 117
  Heat Stroke/Exhaustion 36, 49, 55, 117
  Hyperbaric Shock 53, 68
  Hypothermia 21, 22, 49, 52 - 54, 115
  Recovery Position 51
  Secondary Drowning 24, 51
  Shock 111
  Sun Burn 3, 36, 74, 77
First Aid Kit 15, 46, 49 - 55, 71, 74, 91, 100
Food 4, 36, 37, 72, 84, 86 - 88, 93, 100, 106

Global Positioning System (GPS) 23, 60

Helicopter 15, 17, 28, 66, 67, 68

Igloo 103
Impact 15, 17
Infant Life Preserver 22
Inflating Liferafts 31
Insects 1, 70, 90, 91, 92 - 93, 111

Knives 34

Lashing 85
Lavatory Arrangements 88, 96, 105
Leadership 2, 5
Lifejacket 8, 14, 15, 22, 24, 28, 29, 52, 53, 67, 99
Lifejacket Spray Hood 29, 52
Lifeline 44, 93
Liferaft 3, 8, 9, 12, 21, 27, 31, 49, 56, 67, 77
Liferaft Ballast Pockets 43
Liferaft Canopy 39, 55, 67
Liferaft Design 38 - 48
Liferaft Freeboard 42
Liferaft Inflatable Floor 41, 55
Liferaft Look-Out 33, 35
Liferaft Management 13, 31 - 37

Liferaft Repair Kit 46
Liferaft Repairs 35, 36
Liferaft Shape 38
Location 4, 34, 72, 119

Morale 5, 6, 13, 24, 34, 35, 92

Paddles 46
Panic 1, 2, 5, 14, 19, 71, 92
Pliers 47
Protection 4, 35, 70, 72, 84
Pump 46

Rescue 2, 4, 5, 9, 23, 56, 67, 70, 74, 79,
    86, 90, 91, 119
Righting Capsized Liferafts 32
Roster 3, 34, 36, 92
Royal Air Force 10, 18

Safety & Emergency Procedures (SEP) 20
Salt 79
Satellites 59, 60
Seat Belts 15, 19
Seat Cushions 26
Sewing 85
Sharks 30, 35
Shelters 4, 21, 70, 72, 74, 76, 78, 84 - 86,
    93 - 95, 100 - 105, 111, 119
Signalling 15, 40 - 47, 56 - 62
Signalling Equipment
Emergency Locator Transmitter (ELT) 44,
    57 - 60, 90, 120
  Flares

    Hand-Held 46, 64, 120
    Parachute 46, 63, 120
  Sea Dye Marker 47, 63, 121
  SEE/RESCUE Streamer 64, 65, 123
  Signalling Mirror 47, 62, 63, 121
  Strobe 65, 122
  Transponder 56, 120
  Whistle 64, 121
Snow Hole 101, 102
Snow Shoes 99
Snow Trench 102
Solar Still 81
Steven Callahan 3, 66
Stress 3
Survival Blanket 46
Survival Kit 48
Survival Times - water 22, 23

Testing Food 87, 88
Training 6, 7, 14, 19, 25, 30, 40
Training 6, 7, 14, 19, 25, 30, 40
Transpiration Bag 81

Underwater Escape 17 - 20, 23, 29
Uniforms 28

Varig Brazilian Airlines 70, 89, 119

Water 1, 4, 15, 36, 37, 45, 70, 71, 72, 73,
    77, 78, 79 - 84, 86, 91, 93, 100, 105 -
    106
Water Decontamination Tablets 47, 79, 93
Water Makers 36, 37, 45, 93